MANUAL ON PRACTICAL ENTOMOLOGY IN MALARIA

prepared by the
WHO Division of Malaria and Other Parasitic Diseases

PART I
Vector Bionomics and Organization of Anti-Malaria Activities

WORLD HEALTH ORGANIZATION

GENEVA

1975

CONTENTS

PART I. VECTOR BIONOMICS AND ORGANIZATION OF ANTI-MALARIA ACTIVITIES

Page

Preface

Introduction 1

Chapter 1. Vector bionomics and related aspects 3

Chapter 2. Entomology in malaria programmes 55

Chapter 3. Critical appraisal techniques of sampling mosquito populations 101

Chapter 4. Recording and reporting of data 109

Chapter 5. Interpretation of entomological and related data 119

General bibliography 160

PART II. METHODS AND TECHNIQUES

1. Hand collection 1

2. Spray sheet collection 6

3. Catches off bait 12

4. Trap collection 15

5. Experimental huts 39

6. Catches in outdoor shelters 45

7. Larval surveys 55

8. Preservation of material 69

9. Processing the collected samples of a mosquito population 80

10. The study of vector infection and infectivity 112

11. Determination of the duration of the gonotrophic cycle 121

12. Identification of blood meals 127

13. Techniques and procedures for squash preparations of the polytene chromosomes . . 135

14. Vector susceptibility to insecticides 140

15. Bio-assay test 152

Contents (continued)

PART II (continued)

	Page
16. Behavioural reaction of vectors to insecticides	157
17. Physiological test for the tolerance of anopheline larvae to salinity	162
18. Determination of salinity	163
19. Breeding and maintenance of mosquitos under laboratory conditions	165
20. Mosquito rearing under field conditions	173
21. Artificial mating	175
22. Methods for marking mosquitos	179
Bibliography	186

PART I

VECTOR BIONOMICS AND ORGANIZATION OF ANTI-MALARIA ACTIVITIES

Preface

The present manual is intended to be a practical vade-mecum for the entomologist as well as for the malariologist in a malaria programme. For this, in addition to specific entomological procedures, it will be indicated why, when and where regular and special entomological activities need to be carried out and how the data should be analysed for a better overall understanding of the malaria situation.

In revising "Practical Entomology in Malaria Eradication" (1963) a broader treatment has been given to the subject matter as is indicated in the change of title. For one thing consideration is of course given to vector bionomics but, more important, entomology is dealt with as an integral part of malaria epidemiology and not only the description of techniques, collection and tabulation of data confined to vectors alone. This is essential because the entomologist must take an interest in all aspects of malaria as a member of a team which provides an epidemiological service. The information he supplies should always indicate the appropriate action that needs to be taken. Thus, the entomologist has an important role in the planning of a malaria control programme in addition to the more specialized planning of the specific activities of his own service.

The recording, presentation and interpretation of data and the preparation of reports are essential factors for stimulating appropriate action, and examples of reporting forms are provided.

It is also considered that the manual will be useful not only as a guide for entomologists but also as teaching reference material which will give other malaria personnel a better understanding of the use and value of entomological procedures in the overall epidemiological appraisal. Part I covers the bionomics of mosquito populations, the organization of anti-malaria activities and the recording, reporting, and interpretation of entomological data. Part II deals with methods and techniques.

The direct and indirect contributions of WHO staff and numerous investigators in various parts of the world are gratefully acknowledged.

INTRODUCTION

In order to put malaria entomology in perspective, it is useful to regard malaria transmission as part of a natural ecological system. The main factors involved in this system are the parasite, the vector and the human host which react with one another and also with their wider biological and physical environment. This system may be represented in a simple form, as shown below, the arrows indicating the direction of possible influences.

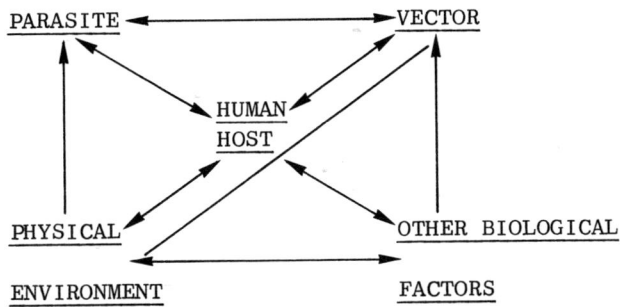

These main factors may be broken down into various components as follows:-

PARASITE

Species
Strain
Temperature requirements for extrinsic cycle

VECTOR

Reproductive requirements
Temperature and humidity requirements
Contact with man
Susceptibility to infection
Feeding and resting behaviour
Flight capabilities
Seasonal distribution
Diapause
Longevity
Ability to develop certain reaction to insecticides

HUMAN HOST

Social organization (urban/rural)
Quality of housing, drainage, water supply, etc.
Occupation
Agricultural pattern (irrigation, etc.)
Population movement (migration, nomadism)
Immunity
Intervention measures (on parasite, mosquito or environment)

PHYSICAL ENVIRONMENT

    Temperature
    Humidity
    Rainfall
    Wind
    Altitude
    Topography
    Water table
    Soil
    Pesticide use

OTHER BIOLOGICAL FACTORS

    Predators
    Parasites
    Pathogens
    Genetic

In the attempt to reduce or interrupt malaria transmission man may alter the physical, chemical or biological components of a given system, for example by drainage, by the use of anti-malarial drugs or insecticides, by the introduction of fish or by genetic manipulation. These measures may result in varying degrees of success or there may be a compensation reaction within the system tending to maintain the status quo, for example the development of resistance of the parasite to drugs, of the vector to insecticides or of the human population to having their houses regularly sprayed.

It is apparent, therefore, that the malaria entomologist requires a high degree of ingenuity in his approach to a given situation and the ability to identify and exploit the most vulnerable links in a complex natural system. It must be stressed that the skill and performance of the members of the entomological team constitute important elements in the application of entomological methods and collection of data. In addition, the entomologist and the malariologist should be aware of the limitations of the entomological sampling techniques, the hazard of drawing conclusions from data of inferior quality and ways and means for improvement of the techniques. This is specially treated in Chapter 3. Therefore, it is sufficient to state here that a well trained and properly oriented and motivated staff supported with adequate facilities is essential for achieving the desired objectives.

Whatever or wherever a malaria problem may be, each factor and aspect has to be considered in the light of the local characteristics. The term "malariogenic potential" describes the capability of the main variables acting together to produce endemic or epidemic malaria in a given locality. For instance, the frequency with which the inhabitants of an area are bitten by infected mosquitos is closely related to the "transmission index" which is the percentage of new cases found among susceptible persons during a given unit of time. The number of infected mosquitos found in a sample is related to the adequacy of the sample taken, to the man-vector contact, to the longevity of the vector(s) and to the number of persons who were infective for the vector. The interaction of the entomological factors is further discussed in Chapter 1.

The control of the disease, either by a direct reduction of the parasite reservoir in man, by reduction of vector longevity or by diminution of man-vector contact is dependent upon the effectiveness of the materials used (drugs, insecticides etc.) and the degree and quality of coverage. The complex relations existing between the numerous factors determining a given epidemiological situation demand that parasitological and entomological observations be correlated in view of their cause-effect interaction. A given epidemiological situation may be the result of natural undisturbed causes, or the result of human intervention in the environment or of various types of control measure. When intervention measures are applied, a third factor related to the quality and quantity of measures is introduced. In order to draw a significant picture from the values of those three factors and from the selected variables calculated from them, the data collected should be statistically as well as qualitatively acceptable and presented in such a way that each technical member of the multi-disciplinary malaria team or service can assess the value of his own measurements, and be able to interpret all data collected by his colleagues in the general framework of a given epidemiological situation.

The approach to a satisfactory understanding of local malaria epidemiology, which changes under the impact of control measures, varies according to the specific intermediate objectives of the programme. Different sampling methods, procedures and criteria are used in the parasitological evaluation; the same applies to entomological evaluation. Therefore activities have to be selected according to the type of criteria and to the data required at the different stages of programme implementation.

The purpose of the baseline entomological observations is to determine the role of the vector mosquitos in the dynamics of transmission, involving the study of their bionomics, behaviour and seasonal prevalence. This, together with the parasitological evidence of season of transmission, will help in the selection of the type of anti-vector measures to be employed, frequency and timing of application. Entomological observations will also assess the effects of control measures on the vectors and enable comparison of achievements with aims pursued. If results are short of the intermediate targets, the comparison of entomological data, including mosquito behaviour and insecticide susceptibility with those revealing the degree of residual transmission and those related to standard of operations, will provide a meaningful indication of the changes needed either in the method of attack or in the performance of operations.

A bibliography relating to each topic will be found at the end of the relevant section.

CHAPTER 1.   VECTOR BIONOMICS AND RELATED ASPECTS

CONTENTS

Page

1. INTRODUCTION 5
2. THE INFLUENCE OF SOME PHYSICAL FACTORS OF THE ENVIRONMENT 7
3. SYNOPSIS OF ECOLOGY OF ADULT MOSQUITO POPULATIONS 9
4. ECOLOGY OF AQUATIC STAGES OF MOSQUITOS WITH SPECIAL REFERENCE TO ANOPHELINE VECTORS OF MALARIA 13
5. THE NATURAL ENEMIES OF MOSQUITOS 17
6. THE INTERACTION OF ENTOMOLOGICAL FACTORS IN MALARIA TRANSMISSION 33
7. VECTOR RESISTANCE TO INSECTICIDES 36
8. CYTOGENETICS IN STUDIES OF MALARIA VECTORS 41

BIBLIOGRAPHY 52

CHAPTER 1

VECTOR BIONOMICS AND RELATED ASPECTS

1. INTRODUCTION

Although this manual is intended to deal primarily with the entomological component in malaria programmes and the techniques to be utilized in studying the bionomics of the vectors, a general outline of the bionomics of mosquito vectors is necessary as a background knowledge for orienting the worker to the importance of adequate application of methods and sound utilization of data.

<u>Definition</u>: Bionomics is that part of biology (often called autecology) which deals with the relationships of a given species and its environment.

Basic studies on the bionomics of mosquitos include the development of immature stages, i.e., eggs, larvae, and pupae as well as the life of the adults under the influences exercised by the environmental conditions.

Behaviour is the result of the interaction of:

- genetic factors, which govern the basic lines of behaviour, and

- ecological factors, which may produce different types of reaction in a population having the same genetic characteristics.

The larval and adult stages of mosquitos exist in two different environments, each stage being under the influence of its immediate surroundings. The adaptation of eggs, larvae, pupae and adults to certain environmental patterns constitutes the influence of factors controlling the seasonal and geographical distribution of the species.

Among the ecological factors, the phenological[1] one is of considerable importance. For example, the ability of mosquitos to breed in fresh or salt water, or both, is controlled by genetic factors. Host preference seems to also be governed by genetic factors, but the intensity of feeding on a certain host may vary from place to place and even from day to day, not only with the availability of hosts, but also with the changes in meteorological conditions. The speed of development of mosquitos is dependent mainly on climatic factors. The speed of the metabolic processes of insects is largely controlled by temperature, hence certain biological events such as the duration of the aquatic stages, the speed of blood digestion and maturation of the ovaries and consequently the frequency of feeding, vary according to temperature.

These features underline the importance of carefully recording the environmental factors at the time of carrying out entomological observations.

It is well known also that under the influence of environmental conditions a vector species may show changes in the seasonal distribution in the same area of dominance. The increase in density of a vector species is very much dependent on climatological factors favourable for its breeding, and adult survival. An exceptionally heavy rainy season might be favourable to the development of a number of species yet detrimental to others.

A mosquito species is regarded as endophilic when largely found resting indoors, or exophilic when largely found resting outdoors, although partial exophily has also been observed. Selection of the resting habitats is much influenced by the temperature and humidity of favourable resting places and human habitats. Endophilic species might exhibit

---

[1] Phenology: study of seasonal periodic biological events.

a considerable degree of exophily when the environmental conditions are favourable. The examples of A. gambiae, A. maculipennis and A. sacharovi under certain ecological conditions are well known. The proportion of A. gambiae resting outdoors is much higher in the rainy season.

To date, a large amount of knowledge has been accumulated about the basic behaviour of the main malaria vectors. Despite this, much is required to fill the gaps in our knowledge, for example, larval survival under field conditions, dispersion of different vector species from the breeding places, the long-range stimuli for orientation to a host, the average duration of the gonotrophic cycle in different seasons, oviposition stimuli that determine the choice of a certain type of breeding place, the factors governing the distribution of mosquitos in the day-time resting sites.

From what has been said above, it will be realized that an understanding of the bionomics of the vector is of key importance in the study of the epidemiology of malaria and in anti-vector measures. To summarize, the study of the vector life cycle and behaviour includes the following main points:

(a) site and time of oviposition and embryonic development;

(b) larval development and key factors influencing instar mortality such as natural enemies;

(c) emergence, adult dispersion, mating, gonotrophic concordance, orientation to a host and host preference, and resting behaviour;

(d) different tropisms as might be involved in the above aspects for vector species under different environmental conditions: phototropism, photoperiodic reaction, thermotropism, chemotropism, reaction to insecticides.

A vector's infectivity depends, amongst other factors, on its bionomics which may constitute the most important element in the chain of transmission. It is well known that in the absence of favourable environmental factors, the vector cannot exist; if it does exist, the temperature might not be favourable for sporogony, or the environmental factors may be such that contact with man is not sufficient to ensure transmission even in those cases where the species is genetically susceptible to infection with malaria parasites.

Planning, execution and evaluation of anti-vector measures have to be based on a perfect knowledge of the bionomics of the vector. One cannot expect a successful programme of residual spraying in an area where the vector exhibits a strong natural or obligatory exophily, or when the conditions are such (e.g., houses with scanty walls) that spraying is not sufficient to interrupt transmission, because sufficient numbers of vectors survive outside and will be capable of maintaining transmission indefinitely. In this respect the example of A. balabacensis is well known.[2] Such knowledge is also necessary for the detection of possible change in vector bionomics under the influence of insecticides.

A knowledge of the breeding, resting and biting habits, and longevity of a vector species is therefore essential for organizing anti-vector measures and the evaluation of the impact of such measures. It is essential to carry out such observations before initiating anti-vector measures. As the above-mentioned aspects differ with the variations in environmental conditions (which are local and seasonal in character) it is easily understood why entomological studies are necessary for all representative situations and that the results cannot be extrapolated from one area to another or from one season to another, unless the environmental conditions are exactly the same.

---

[2] This is an outdoor resting species adapted to the jungle. It feeds in high numbers on human beings living near or in the forest, entering the houses during the night and returning to outside shelters again after having fed.

From all that has been mentioned, it is obvious that data on climatic factors, i.e., temperature, humidity, rainfall, wind, degree of insolation, etc., as well as other ecological factors, are of considerable importance for the study of dynamics of vector populations. Such information should be collected not only in general but at the time and place of observations.

Each species is adapted to live in a particular "niche" in the community.[3] The species is in a figurative sense a "prisoner" in its general habitat due to its adaptive dependence on environmental factors, such as climate and food and type of breeding place. The individuals of each species have instinctive reactions, or tropisms, which enable the mosquitos to find the best conditions for maintenance of the species.

In short, the environmental factors which govern the distribution, abundance and density of mosquitos are climate, physical and chemical conditions of the habitat, hosts, enemies and in some cases competition.

The important environmental factors are briefly discussed and synopses of adult and larval mosquito ecology are given in two sections. For more detailed reading, text-books and specific publications should be consulted.

2. THE INFLUENCE OF SOME PHYSICAL FACTORS OF THE ENVIRONMENT

2.1 MACRO AND MICROCLIMATES

The climate is one of the major components of the physical environment, and is a composite condition of which temperature, relative humidity, precipitation, light and wind are the important components. The daily expression of climate is called "weather" which has a profound impact on the biology, distribution and density of a mosquito species at any given time.

The climate can be divided into two types: (i) macroclimate, which means the average weather conditions of an area, and (ii) microclimate, or modifications in restricted areas within the overall macroclimatic zone.

The conditions of the macroclimate will determine the existence or possible area of distribution of a species, whereas the conditions in the microclimate will influence the local distribution of a species inside the zones with the same macroclimate. The microclimate is a modification to some extent of the conditions of the macroclimate. A difference in temperature and humidity of some degrees may exist between the macro- and the microclimate. The mosquitos' resting places have their own microclimate and outdoor resting places have in general a different microclimate from the indoor resting places. When the microclimate of the indoor and outdoor resting places is similar, partially endophilic mosquitos will also be found in high numbers in the outside resting places but will practically disappear from outside when the outside microclimate changes, as occurs in areas with periodic dry seasons. The macroclimate has an important influence on the microclimate of different sites, e.g., the temperature inside a house might be a few degrees (1-2°C) less than the outdoor temperature, whereas the relative humidity might be much higher than outdoors, being sometimes 20-30% higher than the outdoor humidity. The daily increase or decrease of temperature inside houses usually follows the outdoor temperature, one or two hours later in the absence of artificial heating. During the night the temperature inside is higher in the first hours of the evening and lower during the first hours after sunrise than the outdoor temperature. Type of construction, materials and ventilation have a profound influence on the changes in the microclimate of the indoor resting sites.

---

[3] Community can be defined as the biological component of a natural area. Each community has a definite set of species which live in a vital relationship.

The macroclimate varies according to the latitude and altitude and the prevailing meteorological factors, but the microclimate varies with the presence and absence of different types of vegetation, exposure, aspect, colour, and other peculiarities of a given site which produce a difference in the conditions of temperature and humidity, when compared with the macroclimate.

2.2 TEMPERATURE

Insects are cold-blooded and therefore all metabolic processes and the entire vital cycle depends on the environmental temperature. Mosquitos, like the majority of insects, are unable to control the temperature of their bodies to any great extent. The insects can survive low temperatures but their metabolic processes are slowed down or even arrested when temperature falls below the threshold. At temperatures higher than 32-35°C the metabolism is also modified in the sense of slowing the physiological process. The average optimum temperature for the development of most mosquito species is around 25-27°C. The development can be completely arrested at 10°C or over 40°C when a high mortality may occur. The tolerance to temperature depends on the species and in general a species will not support for long an increase in temperature of the environment 5-6°C over the limits to which the species is normally adapted.

The gonotrophic cycle, including digestion of blood, depends on the temperature.

The influence on the longevity of insects is also very strong. Tropical species will not withstand temperatures near freezing point. Permanent high temperature over 27-30°C will reduce the average life of a mosquito population.

A. maculipennis larvae (a temperate zone species) will become completely inactive, remain at the surface and will be killed during the freezing of the superficial layer of the water, whereas larvae of A. claviger or A. plumbeus will move actively at temperatures around 0°C and survive in pockets of water included in ice.

In areas where there is a winter period, one stage of the vital cycle is adapted to survive at low temperatures. A. maculipennis larvae are killed at 0°C but the adults can survive temperatures of -20°C. A. claviger and A. plumbeus are better adapted to survive the winter as larvae than as adults. The mosquitos are adapted to their normal environment, with the severity of the seasons of hot and cold weather playing a role as factors restricting their distribution.

2.3 HUMIDITY

This can also act as a limiting factor in distribution and longevity. Owing to the tracheal system of respiration, insects in general are particularly susceptible to dessication. Forest species are more susceptible to humidity changes than those living in areas with a dry climate. During dry weather, indoor mosquitos will concentrate in those houses or other indoor resting places where the microclimate offers a favourable humidity. Outdoor-resting mosquitos will rest in the vegetation near the ground during dry periods.

During day and night, there is a daily rhythm of temperature and humidity, characteristic for each area. The activity of insects is correlated with this rhythm except for those in diapause. Such a rhythm is most evident in areas with hot and relatively dry days. During the day mosquitos rest inactive in cooler places. At dusk, when the temperatures drop and the humidity increases, the mosquitos suddenly become activated.

2.4 RAINFALL

The effect of rainfall varies according to its amount and the physical features of the terrain. Repeated rains cause severe flooding, resulting in temporary flushing out of the breeding places. Consequently, the breeding of a vector population is greatly reduced but it will soon be re-established when normal conditions are restored. Moderately frequent rainfall but with fairly long periods of sunshine will increase the opportunity for prolific breeding.

2.5 LIGHT

Although circadian rhythms have been found to play an important role in controlling most mosquito activities, the effect of light can generally be correlated with movement of mosquitos for feeding or resting. Most anopheline mosquitos are crepuscular and night feeders. Some species such as Anopheles cruzi and A. bellator bite during day and night. Endophilic species such as A. gambiae and A. stephensi under normal temperature conditions start to leave the houses when the light intensity decreases about 20-30 minutes after sunset. These consist mainly of gravid females leaving to seek a breeding site for oviposition, and partly of hungry females activated to seek a blood meal if a host is not readily available in the houses where they have sheltered. At dawn, mosquitos seek day-time resting shelters, and light intensity together with humidity controls the choice of resting site indoors and outdoors. Mosquito species vary in their preference for certain light intensities in the resting places.

In the temperate zone, the photoperiod plays a major role in inducing hibernation in anopheline mosquitos. With the onset of short-day conditions in early autumn, non-gonoactive females of A. maculipennis destined for hibernation appear.

3. SYNOPSIS OF ECOLOGY OF ADULT MOSQUITO POPULATIONS

3.1 THE HABITAT OF ADULT MOSQUITO POPULATIONS

The adult mosquito's habitat, in a broad sense, can be defined as an area where favourable resting places, hosts and breeding places are present. Factors affecting the choice of the mosquito resting habitat include: temperature, humidity, protection against sunlight, wind, predators, etc.

Each mosquito species orientates during various activities to that part of its normal habitat where favourable conditions exist to satisfy the vital needs of each physiological state. Mosquitos concentrate in such areas, flying sometimes long distances from the breeding sites in search of food and consequently favourable resting sites.

The movements of mosquito populations inside different parts of their habitat are governed by a number of factors, such as temperature, humidity, host attractiveness and the attractiveness of breeding sites. The influence of such factors is dependent on the physiological condition of the mosquitos (e.g. the attraction of the host has less influence on fed mosquitos). The flight to breeding places is stimulated by the full development of the ovaries and gravid females will be attracted by particular characteristics of specific breeding sites. Adverse temperature and humidity and absence of host will make mosquitos change their resting place; only non-inseminated females will be attracted to a male swarm. The movements of mosquito populations in general are in the direction of (i) post emergence or oviposition resting, (ii) feeding, (iii) day-time resting, and (iv) breeding sites.

3.2 EMERGENCE OF MOSQUITOS

It appears that the emergence of adult mosquitos follows a given rhythm. Massive emergence occurs during the night but day-time emergence has also been observed. When the emergence of a single batch is observed the males emerge first, reaching peak emergence before the females. In general, at the end of the emergence period, only females remain to emerge from the pupae.

3.3 MATING

Mating usually occurs during 24-48 hours after emergence, before feeding, but a number of females will take a blood meal and will mate afterwards. Some species need a swarming (nuptial dance) of males for mating (eurygamous species), other species do not need a nuptial dance and therefore can also mate in small cages (stenogamous species). Mating occurs, in general, only once in the life of the female mosquito but may occur twice, mainly in the stenogamous species. The inseminated female mosquito is easily recognized by dissecting the spermatheca and examining it for the presence of the spermatozoa. Non-inseminated females normally do not develop their ovaries even when taking several blood meals, but if they do the eggs laid are sterile. Swarming occurs in the area of the breeding places or inside the area where mosquitos are resting, i.e. at the corner of the animal stable, houses, bushes, trees or even over animals at rest or moving. Swarming starts in the crepuscular light and often ends after a short period in complete darkness. It starts again early in the morning at dawn, lasting till before sunrise. The females enter for a few seconds into the swarms and, on being seized by a male, the pair drops out.

3.4 DISPERSION AND RESTING BEHAVIOUR

Dispersion can be considered from two viewpoints: (a) active dispersion when mosquitos reach different areas during their normal flight behaviour and (b) passive dispersion when mosquitos are transported by air currents and/or vehicles (cars, buses, trains, etc.).

After emerging the mosquitos remain for some hours in the vicinity of the breeding site. Some then fly to areas where hosts are available, orientated by host stimuli. Newly emerged mosquitos which have not left the area of the breeding site before sunrise will remain there until the next evening. The same thing happens with parous females who have laid eggs in the early morning. When vegetation is absent around the breeding sites, the mosquitos are forced to leave the area as soon as possible, flying to the nearest favourable human habitation or natural harbourage. Males generally tend to be more concentrated in the area of the breeding site and to remain in outdoor shelters, although a good number of males of the endophilic species accompany the females to their resting places.

A high proportion of males will generally indicate the proximity of breeding sites. Generally speaking, the flight range for a number of malaria vectors is about 1-3 kilometres, although some mosquito females will fly longer distances. However, species differ in their flight range. Movement of cattle from the breeding place area in the evening to remote villages may lead mosquitos in flight to distances of over 10 kilometres. Wind causes dispersion over a wide range. For example, Anopheles pharoensis was reported to have appeared in the desert in Egypt at a distance of 56 km from the nearest possible breeding places and another record shows its appearance at a camp in the western desert at a distance of 29 km and from the nearest surface water. The sudden appearance of A. pharoensis attacking man was in the middle of a single night in each of two consecutive months. On the other hand, under conditions of winds of very low speeds, there is some evidence that mosquitos can detect the air-carried host-specific odours from a distance and orient themselves to the host, flying upwind.

The areas bordering the breeding sites have the highest density in all situations. Some species, such as A. gambiae, are more strictly concentrated, during the dry season, near the breeding area than are A. funestus.

There is a continual movement of female mosquitos from breeding sites to resting places, to the host and then back to the resting places and then once more to the breeding sites (Fig. 1). Such cyclic movements will be repeated during the whole life of the adult female. After emergence a female mosquito may fly to a village situated a good distance away from the breeding site but, after the maturation of the ovaries, it will have to return to the breeding site area and again to the area where hosts are present, possibly entering other places and feeding on different hosts. A mosquito might become infected in a house near the breeding site and then might reach part of the village or even another village far away from the breeding site by the time it becomes infective, thus transmitting the infection there. When the breeding sites are situated in an area between two villages infected gravid mosquitos from one village going to the breeding places may fly after oviposition to the other village, and may actively or passively reach the other village effecting transmission of infection.

- 11 -

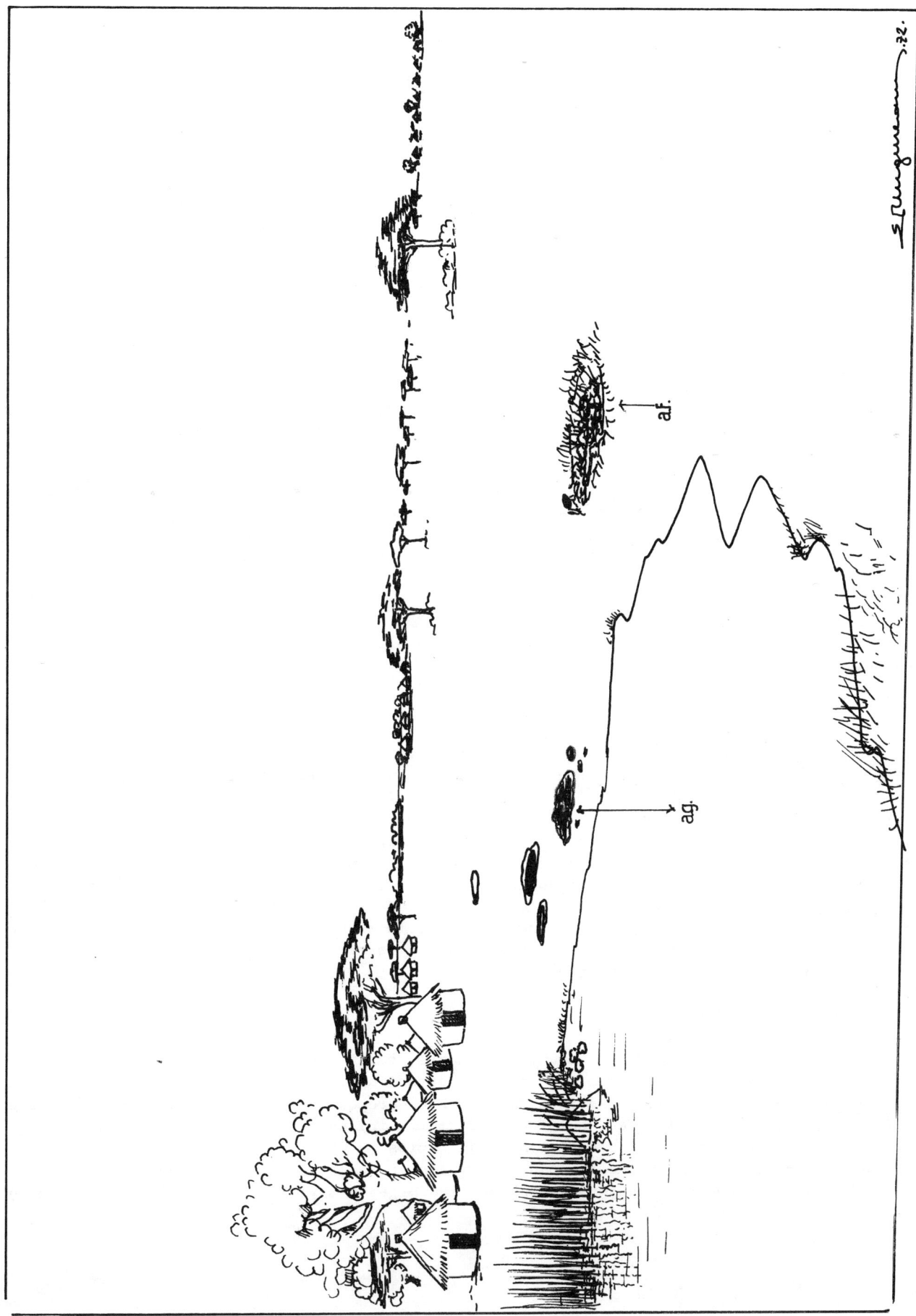

Fig. 1. Larval sites of Anopheles gambiae (ag) and A. funestus (af).

Such a situation has been reported from neighbouring countries where the breeding site is situated on a frontier or just on one side of a frontier; infected mosquitos from one country might introduce the disease to neighbouring areas in the other country. The study of dispersion of different vectors is important in this context and also in relation to protection of a circumscribed area under malaria control by preventing the infiltration of vectors from the surrounding unprotected areas.

Mosquitos will concentrate during the day in places where the optimum resting conditions for each species are met. Endophilic mosquitos are usually found inside the house during the day-time; they may attempt to survive outside of the indoor environment, in spots with favourable microclimate, but the proportion that does so is particularly small. The movement of indoor-resting mosquitos in search of food and a resting place results in them concentrating and resting in the proximity of the host on which they feed.

Individuals of species with exophilic tendencies can be found indoors when the indoor conditions are those required by these species, but the number of those which can rest indoors is much lower than the number resting normally outdoors. Mosquito species which are exclusively exophilic may be found indoors only for short periods before and after feeding during the night. The concentration of exophilic mosquitos at ground level or on the roots of trees is due to the need for humidity and shade during the day. Such concentrations usually occur in the proximity of favourable breeding sites and hosts but when favourable resting sites are numerous, mosquitos disperse over a large area, hence the difficulty of finding adequate numbers during the day. Males will concentrate far away from the host, the majority in the breeding site area. Many exophilic species will concentrate, temporarily, during the night around the host, before and/or after feeding, for variable periods of time.

3.5 FEEDING BEHAVIOUR

This is considered in relation to the time, the place, the host and the frequency of feeding.

3.5.1 Time of feeding

Anopheles species, with very few exceptions, have a crepuscular and nocturnal activity; during the day-time they rest in favourable resting places. Flight activities are performed in a daily rhythm which is governed by climatological factors (temperature and humidity) and physiological needs, such as mating, feeding, resting and egg laying. With regard to biting, some species have intense activity early in the evening, others the whole night, increasing after 22.00 hours. The majority of species have two peaks of activity: the first, the highest peak of activity, is before midnight and the second, at dawn; these can change with the humidity, temperature and wind which might increase or decrease the biting activity. Exophilic forest mosquitos will fly during the day-time in search of a more favourable source of food. The light, humidity and temperature in the forest are in some ways similar to those existing during twilight in open areas.

3.5.2 The place of feeding

Exophilic mosquitos feed predominantly outdoors, but a high proportion of some species will feed indoors when humans become the most important and preferable hosts (e.g., A. sinensis, A. balabacensis). Endophilic mosquitos feed in high proportions inside, but, when hosts are available, a good number also feed outside. Climatic influences, such as temperature, humidity, rain, etc., have an influence on human habits as well as on mosquitos. It is important, therefore, when estimating man-vector contact that the distribution of the human population between indoor and outdoor sites and the night activities of certain groups and in certain seasons should be carefully studied.

3.5.3 The host

Mosquitos have been classified, as far as feeding preferences on different hosts are concerned, into three main categories:

(a) anthropophilic mosquitos (preference for feeding on man);

(b) zoophilic mosquitos (preference for feeding on animals);

(c) without fixed preference for one group or another, i.e., indiscriminate biters.

However, innate host preference is difficult to demonstrate. Genetic differences do exist but there is no doubt that in many cases host <u>availability</u> plays a major or even predominant role. "Host choice" is, therefore, referred to as "host preference" as a general descriptive term.

In the absence of a preferred host, mosquitos will feed on the next best available host. The forage ratio has been employed to express the host preference <u>versus</u> availability of different types of host.

Several factors have been incriminated in orienting the mosquito to a host, such as carbon dioxide, odour, heat, moisture and visual factors.

Recent observations suggested that carbon dioxide is a non-specific factor common to all hosts activating hungry mosquitos and that some host-specific odours carried by a stream of air would enable a mosquito species to orient to a certain host at fairly long distances. However, these aspects are awaiting further investigations. Similarly, human beings exhibit variation in attractiveness to mosquito bite. Factors such as body temperature, sweat and others may play a role but conclusive evidence is needed.

3.5.4 The frequency of feeding

The frequency of feeding in the case of gonotrophic concordance will depend on the duration of the gonotrophic cycle which has been defined by Detinova (1963) as including the duration of digestion of blood and ovary development, the duration of reaching a suitable breeding place to lay eggs, and the duration of time until feeding again. Unfertilized females will feed at shorter intervals after the ingestion of blood (since they usually do not develop ovaries). A proportion of nulliparous females may take another blood meal one or two days after the digestion of the first meal. These were termed by Gillies (1965) as "pre-gravid" females. During dry periods, gravid females may feed again without laying eggs, as has been frequently observed in <u>A. gambiae</u> by several workers. Hibernating females may have a blood meal at very long intervals (several weeks or months, depending on the species and the region).

4. ECOLOGY OF AQUATIC STAGES OF MOSQUITOS WITH SPECIAL REFERENCE TO ANOPHELINE VECTORS OF MALARIA

4.1 PHYSIOCHEMICAL CONDITIONS OF BREEDING PLACES

Anopheline mosquito larvae live in the surface layer of the water in which the eggs were deposited, concentrating at sites where they find food and a degree or protection mainly against water currents, waves and predators. Their respiratory mechanism is adapted for life at the surface, the spiracles of the tracheal system being in contact with the atmosphere. Some species can also utilize dissolved oxygen for cutaneous respiration. The larvae may leave the surface temporarily for feeding purposes or as an escape reaction. Some species of culicine mosquitos might remain normally for long periods under the surface, where they will feed but they must rise to the surface for respiration.

In contrast, <u>Mansonia</u> larvae have a particular adaptation to obtain the necessary oxygen from the plant tissues. The larvae and pupae live attached by their modified syphon to the roots of aquatic plants, the pupae only come to the surface when the adult form is ready to emerge.

No mosquito larvae are to be found on the open surface of lakes, ponds or rivers but are always found in sheltered niches associated with lakes, streams and marshes, on the borders of water collections or in patches of vegetation either marginal or at some distance from the shore. Temporary rain pools, puddles, and water accumulated in plants (leaf bases, flowers, tree holes) are favourable breeding places for some mosquito species. Some mosquito species breed in clean fresh water whereas others are adapted to breeding in brackish water (A. sundaicus) or highly polluted water with organic matter (A. stephensi) in some areas. Some species are restricted to a single type of breeding habitat while others possess a larger adaptability, but their presence in a given type of breeding place is due to the oviposition habits of the female mosquito, which is the main determining factor of the presence of larvae in different types of larval habitats.

The selection of larval habitat is made by the female adult mosquito, and the preference for one type or types of breeding habitat or another is more or less genetically fixed by natural selection. The same breeding place might attract one species and deter another. The example of Culex fatigans breeding in highly polluted water, and in general the absence of anopheline larvae, is explained by the positive and negative attraction exercised by the same breeding place. Difference in temperature would make a breeding place more attractive to one species than another. A. claviger will lay eggs in breeding places with cool, clear, spring water, and will not be attracted by stagnant breeding places that are exposed to light and have a higher temperature. When the natural breeding places are missing, A. claviger will breed in domestic water cisterns (Syria). A. plumbeus and Aedes geniculatus are found only in tree holes.

Various species of plants are indicators for a given type of water, the plants reflecting not only the physical characteristics of a breeding habitat but also the chemical content and temperature. The absence of vegetation or the presence of a given type of vegetation might suggest the possibility of the presence or absence of a given species of mosquito larvae, in a given type of water (stagnant, running, etc.). A classical example is constituted by the Mansonia larvae which are attached to the roots of floating plants (Pistia, Eichornia). The presence of Bromeliaceae will suggest the possible presence of Kerteszia species or Aedes simpsoni, or A. philippinesis which typically breed in the water accumulated between leaf bases of the plants. The presence of a thick carpet of Lemna or of microscopic algae, e.g. Volvox, covering the surface of stagnant ponds will indicate the absence of anopheline larvae on that surface. In African conditions, a small borrow pit covered with vegetation like Pistia will suggest the presence of A. funestus but not of A. gambiae.

4.2. CLASSIFICATION OF THE LARVAL HABITATS

As shown below, an attempt has been made to classify larval habitats according to size, degree of persistence of the breeding water (permanent or temporary) and type of water designated (modified after Bates, 1941).

A. <u>Large or medium size habitats</u>

<u>Permanent or semi-permanent standing water</u> (fresh or brackish water)

  Marshes or marshy shores of lakes

  Swamps

  Ponds

  Pools

  Seepages

  Reservoirs

  Wells

Running water

    Springs

    Streams

    Rivers

    Canals

Temporary breeding places

    Natural

        Rain pools

        Pools on river banks created by inundation

        Puddles

    Artificial

        Small irrigation water channels

        Irrigated surfaces, rice fields

        Surface ditches for the evacuation of waste water or for drainage of temporary waters

        Footprints

        Stagnant waste water

        Borrow pits

B.   Small well defined habitats

    Container habitats

        Natural

            Tree holes

            Rock holes

            Plant leaves or stems which accumulate water

            Crab holes

        Artificial (man-made)

            Water cisterns

            Earth pots for storage of water

            Flower vases

            Bird fountains

            Empty tins, coconut husks, etc.

            Rubber tyres, etc.

A perfect classification of breeding places is impossible. In this classification we have considered the persistence and the main physical aspect of the water collection. Some of the types mentioned under temporary, artificial, breeding places might also be permanent or semi-permanent but these have been mentioned here only under the heading which in general reflects the characteristics most frequently encountered; when some of the breeding habitats mentioned are permanent or semi-permanent, they should be recorded in this group, as might be the case with irrigation channels; some streams might be only temporary.

4.3 INFORMATION TO BE COLLECTED DURING LARVAL SURVEYS

It is understood that while collecting the data on the breeding places, the following additional information should also be recorded:

Physical aspects

Surface area (or length and width)

Approximate depth

Speed (if running water)

Whether the breeding place is shaded or unshaded

If breeding place becomes completely dry during the dry season

(e.g., temporary collections, small or semi-permanent rivers).

Chemical aspects

Fresh or brackish water

The presence of pollution with organic matter.

Vegetation

Main type of vegetation

- in the breeding place (floating, vertical, mention predominant species if possible)
- around the breeding place (species predominating)

Density of vegetation (compact, dispersed).

General

Use to which the water is put by human inhabitants

General location of breeding places (e.g. in the open, in forest areas, etc.).

4.4 THE DISTRIBUTION OF LARVAE WITHIN THE BREEDING SITE

The distribution of larvae inside the breeding site is not uniform; even in small breeding sites the larvae will always be concentrated around the borders and around different floating objects or vegetation (Spyrogira, etc.). Here the larvae are seeking some support to avoid movement during feeding or to avoid being displaced by the movements of the superficial layer of the water if there is a slight current; the microscopic flora and fauna upon which the larvae feed is more abundant around the vegetation, e.g., as in rice plantations or over the carpets of vegetation submerged a few millimetres below the water surface, depending on the vector species. The perimeter of such favourable areas of development is important for calculating the real surface to be treated with larvicides. The area of a lake free of vegetation, apart from the area near the border two-three metres wide, hardly enters into this calculation of surfaces to come under larviciding because the area is free from larvae.

In a breeding place where the water moves continuously or temporarily, the larvae concentrate in those places where they will find a still harbourage; in a large stagnant or semi-stagnant water collection, first stage larvae are concentrated in places where the eggs were layed, but third and fourth stage larvae move several metres away from the place of hatching and concentrate in the most favourable spots, in the shade or in the light, near vegetation etc., according to the species' habit. Natural or artificial agitation of the water favours such dissemination of larvae, winds moving the superficial layer of the water will force the larvae to concentrate in quiet corners of the breeding place. The water moving inside rice paddies may also produce concentrations of larvae in higher densities at some less affected sites. Rain might wash out some breeding places or cause flooding, resulting in the larvae being transferred long distances, varying from a few metres to several kilometres. Eggs of anophelines may be transported by fast flowing waters and may be destroyed by waves sinking them or throwing them on to dry ground where they are rapidly dried by the sun. When larvae are in an advanced stage of development they are more resistant to dessication and can survive several days on mud in the shade. These few examples show how the environmental conditions may play an important role in determining the concentration of larvae and why they should therefore be carefully considered when estimating larval density.

5. THE NATURAL ENEMIES OF MOSQUITOS

There are numerous records of natural enemies of mosquitos involving invertebrate and vertebrate predators, viruses, bacteria, protozoa, helminths and fungi. The natural enemies of mosquitos, together with other regulating factors, have an important role in the natural balance of fauna by preventing a population explosion. It could be calculated theoretically that from a single female laying 200 eggs (from which 100 females will be produced) there will be produced in the fourth generation about 100 million female mosquitos. The exact role of the natural enemies is difficult to assess. Considerable laboratory observations on the effect of such biological agents have been conducted and some field trials have been attempted. The most success has been achieved with larvivorous fish.

The efficacy of natural enemies is higher when the biotope and the behaviour of larvae and predators favours close contact, for example, <u>Gambusia</u> fish are top feeders as are mosquito larvae, therefore the area of activity of the predator and the prey is the same. Predators which live at the bottom of breeding sites are less efficient. The smaller the volume of the water in which the natural enemies and mosquitos live, the more efficient the predator activity. The presence of other insects on which the predator feeds reduces the impact of the predators on the mosquito population. Also, the efficacy of the predator varies with the environment which might increase or decrease the possibility and duration of contact between predators and mosquitos.

A small volume of water containing pathogens will favour the infection of mosquito larvae. The speed of development of larval stages also has some influence on the efficiency of some infective agents and predators. The longer the aquatic life cycle, the higher the risk of becoming infected or captured. Due consideration should be given to any possible adverse effect of the biological agent to be introduced in the environment on man and domestic animals.

5.1 PREDATORS OF ADULT MOSQUITOS

<u>Insects</u>. The Anthomyid flies, <u>Lispa</u> and <u>Gerris</u>, (pond skaters) hunt adult mosquitos near or on the surface of breeding sites.

Dragonfly adults are strong hunters of mosquito adults in the breeding sites, usually early in the evening at sunset and early in the morning when the main movements of mosquitos occur.

Ceratopogonine midges of the genus <u>Culicoides</u> have been observed attacking engorged mosquitos.

- 18 -

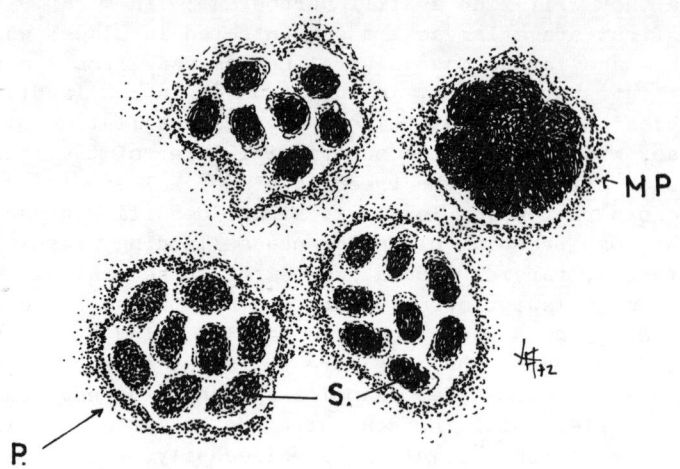

Thelohania legeri

P = parablast with mature spores (S)

MP = maturing parablast

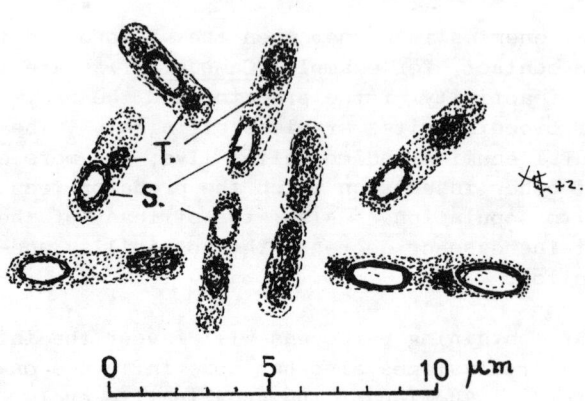

Bacillus thuringiensis

S = spores

T = inclusions of toxin

Fig. 2.   Pathogens of mosquito larvae.

- 19 -

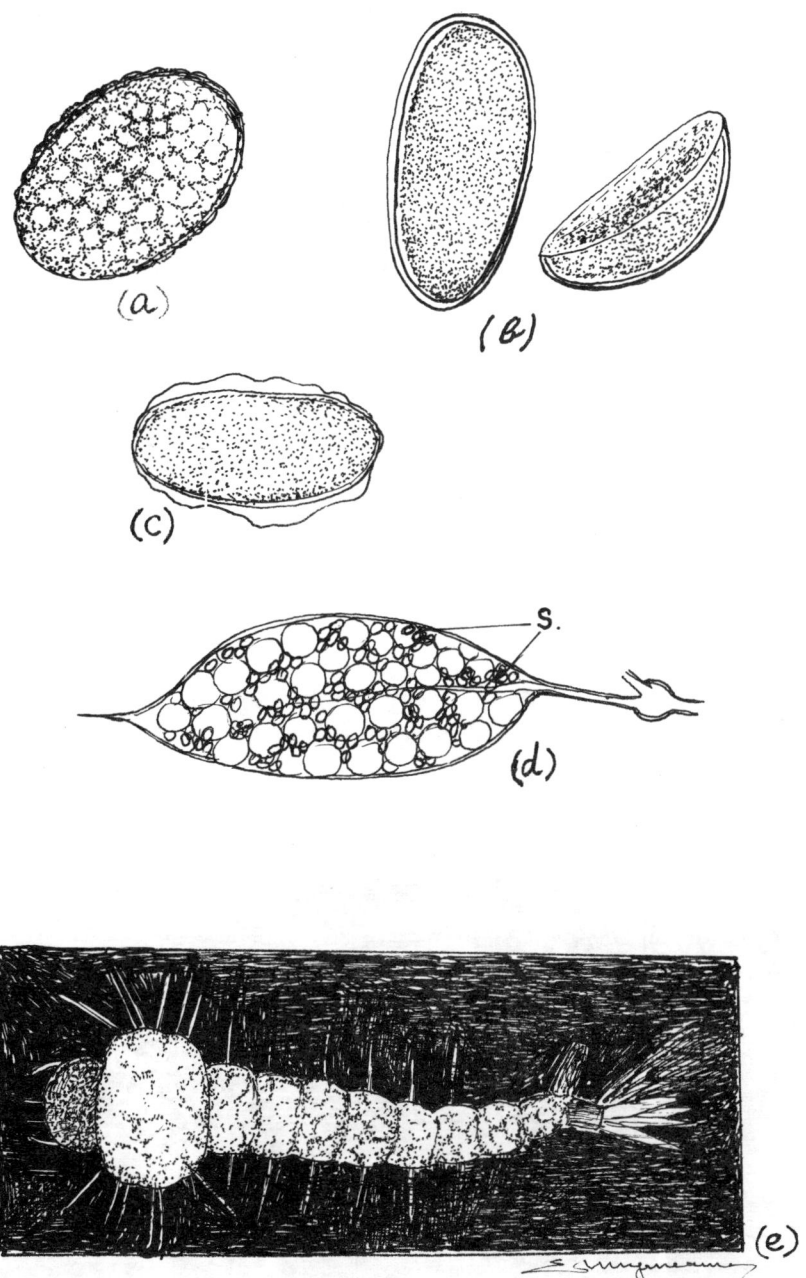

Fig. 3. Pathogens of mosquito larvae: (a) <u>Coelomomyces chironomi</u>; (b) <u>Coelomomyces stegomiae</u>; (c) <u>Coelomomyces raffaeli</u>; (d) Anopheline ovary infected with <u>Coelomomyces</u>, S = sporangia; (e) Larva infected with <u>Thelohania opacita</u>. Infected larvae have a whitish-yellow colour.

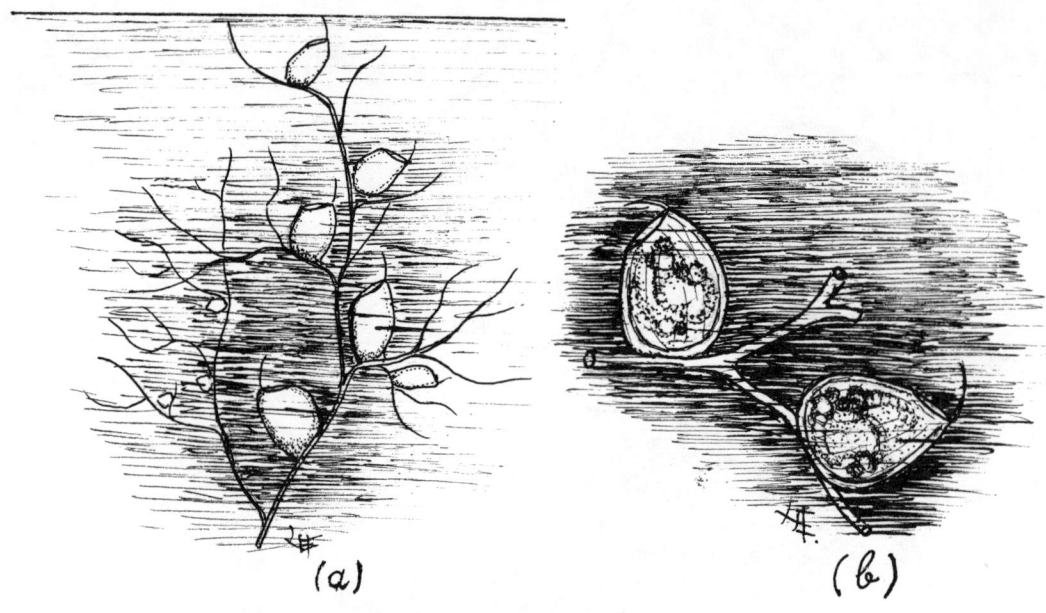

Fig. 4. Predators of mosquito larvae: (a) and (b) Utricularia;
(c) Hydra.

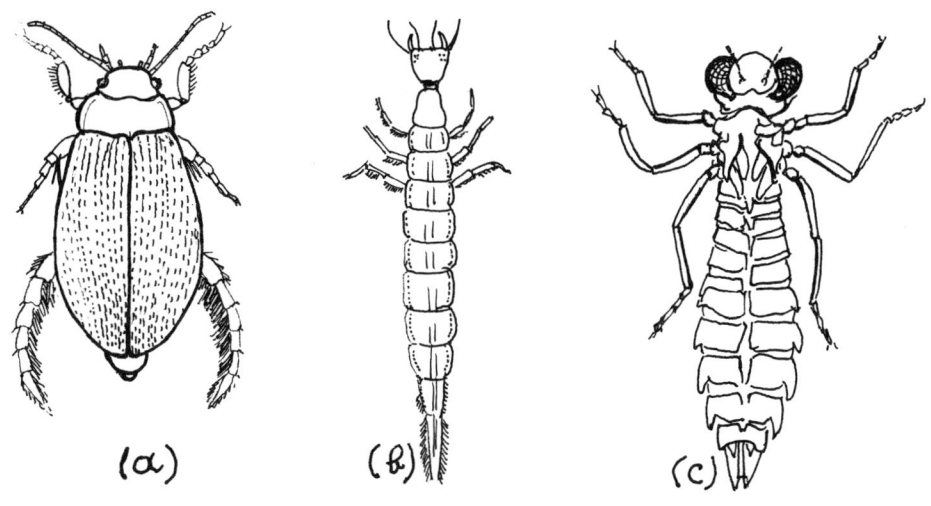

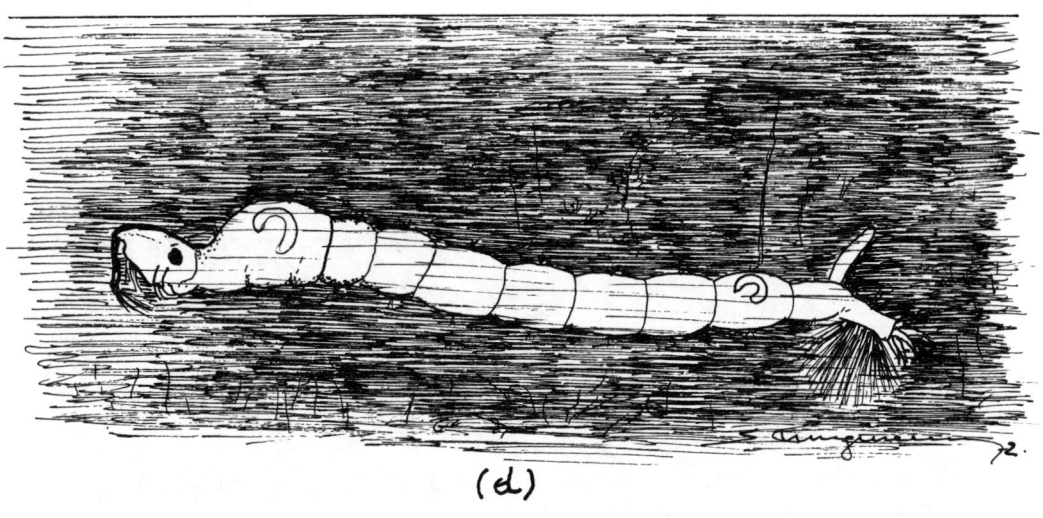

Fig. 5. Some predators of mosquito larvae: (a) <u>Cybister</u> spp., adult; (b) <u>Cybister</u> spp., larva; (c) Dragonfly nymph; (d) <u>Chaoborus</u> spp., larva.

Spiders destroy a number of adult mosquitos, mainly in areas with vegetation where the spiders can suspend their web and inside houses and stables.

Mites. Adult mosquitos are frequently parasitized by larvae of hydracarine mites (water mites) breeding in stagnant water (Fig. 7). In general, adult females showing hydracarine larvae are nulliparous. The adult hydracarines are not found on mosquito adults but they are found in the breeding places. The hydracarine larvae attach themselves to and feed on the adult mosquitos at the moment of emergence from the pupae.

Lizards. In tropical areas, geckos destroy a large number of mosquitos.

Birds. Among the insectivorous birds, swallows take mosquitos in large numbers.

Mammals. Bats hunt insects which fly in the evening. Some decades ago, the possibility was envisaged of using bats as a means to reduce mosquito density. However, this idea did not materialize since bats will feed on any type of flying insect and, therefore, their efficiency in reducing the mosquito population is in practice negligible.

The natural enemies of adult mosquitos unfortunately cannot be manipulated like some of the efficient predators of mosquito larvae (larvivorous fish).

5.2 PATHOGENS AND PARASITES OF LARVAE

Some bacteria cause death or damage to the larvae by the toxins they produce or by forming a scum on the surface of the water. Bacillus thuringiensis is given as an example of pathogenic bacteria (Fig. 2). Leptothrix buccalis infects larvae and is passed into the pupae killing the adults soon after emergence.

Of the microsporidia, Thelohania (Fig. 2) causes damage to the fat bodies and death of larvae occurs at the time of pupation.

With the fungus Coelomomyces (Fig. 3), infected larvae usually die before pupation. If infected adults emerge from infected larvae, ovarian development is prevented. Of the nematodes, Mermithidae parasitize young larvae and causes death. If, as in some species, the worm is passed to the pupa, the adult is killed.

5.3 PREDATORS OF LARVAE

Aquatic plants. Perhaps the best known of these is Utricularia, which traps mosquito larvae as well as other small aquatic animals (Fig. 4).

Coelenterates. Fresh water Hydra can destroy mainly first and second stage larvae of mosquitos breeding in clear, stagnant, cold water with submerged vegetation reaching the surface of the water (Fig. 4).

Insects. (Predatory aquatic insects) The larvae of Dytiscidae and Hydrophilidae (Coleoptera) are redoubtable enemies of mosquitos (Fig. 5). Dragonfly larvae (Fig. 5) also feed on mosquito larvae but seem to be less efficient than the larvae of aquatic Coleoptera. Chaoborus larvae (Fig. 5) also prey upon mosquito larvae.

Larvae of Culex tigripes (Africa), Culex halifaxii (Solomon Islands), Toxorhynchites and Aedes are mentioned as preying on anopheline larvae. When anopheline larvae are extremely crowded in very small breeding places, cannibalism occurs and fourth stage larvae may eat the larvae of the same species or prey on young larvae of other anopheline species.

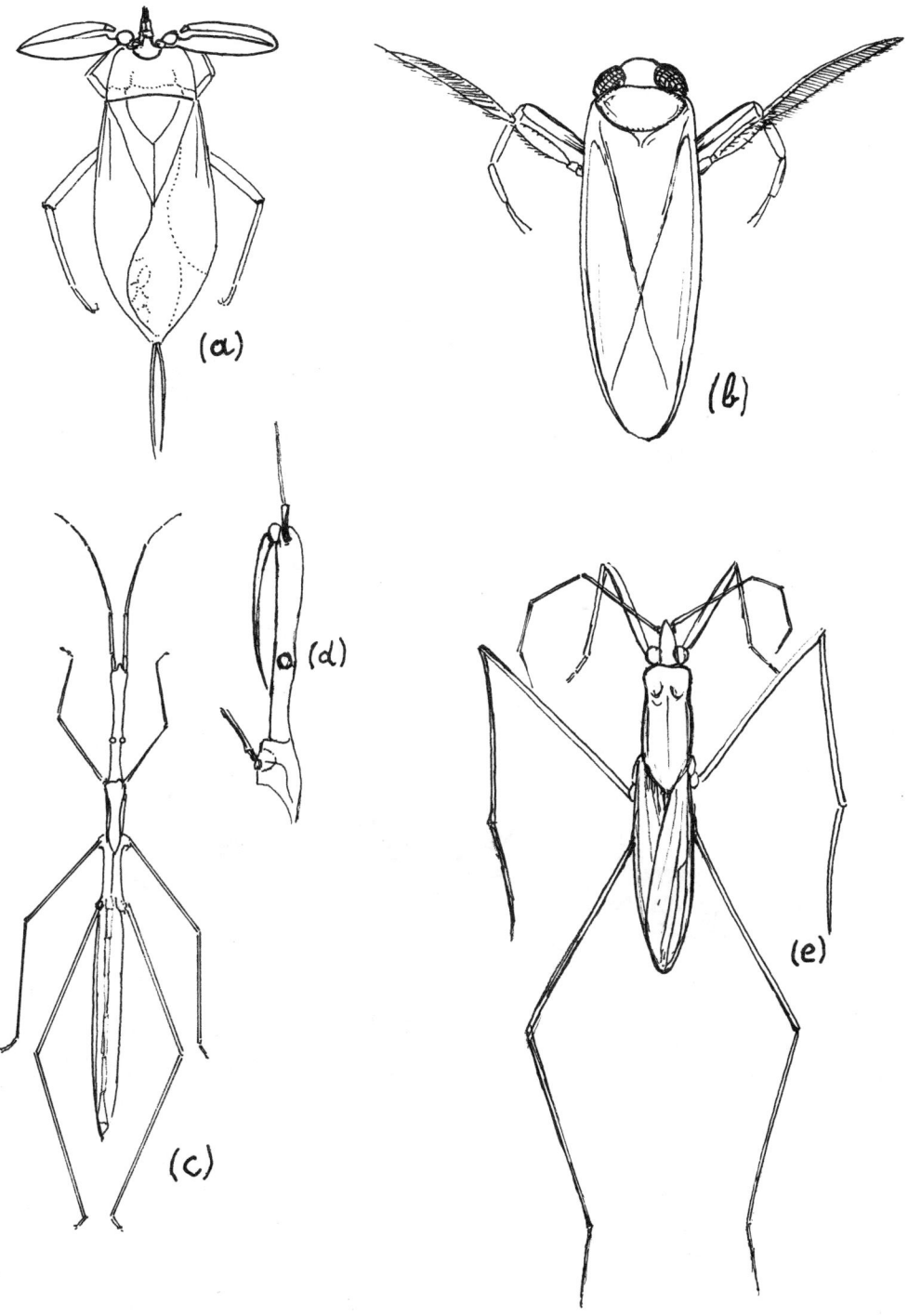

Fig. 6. Predators of mosquito larvae: (a) *Nepa*; (b) *Notonecta*; (c) *Hydrometra*; (d) *Hydrometra*, side view of head showing proboscis; (e) *Gerris*.

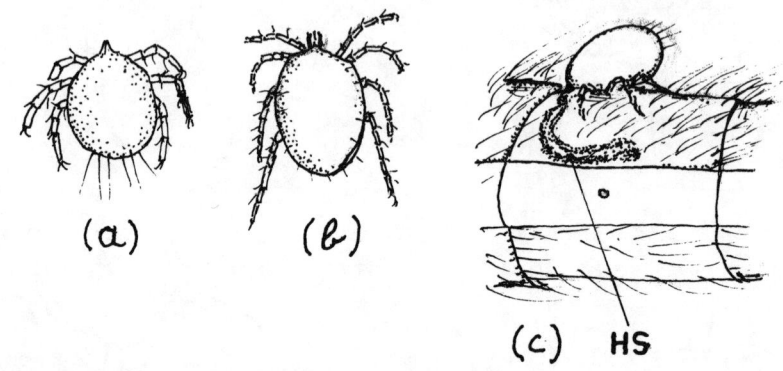
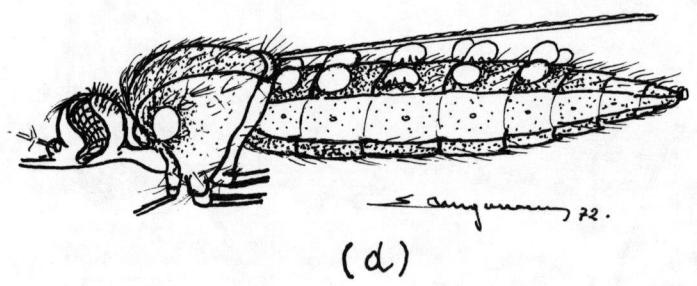

Fig. 7. Parasites of mosquito adults: hydracarine mites - (a) larval mite;
(b) adult mite; (c) and (d) larval mites on body of mosquito
(HS = mouthparts inserted).

Aquatic species of Hemiptera (Fig. 6) such as Nepa (water scorpion), Notonecta (water boatman), Hydrometra (water stick), and Belostoma (giant water bug) prey on mosquito larvae (mainly third and fourth stage) by piercing their bodies with the proboscis and sucking the body fluid.

Gerris (pond skater) (Fig. 6) preys on anopheline larvae as well as on adults. A precipitin test has recently been developed to detect the presence of A. gambiae larvae in guts or squashes of possible predators collected from breeding places, and more studies in this direction would throw light on the effect of predators in regulating populations of anopheline vectors of malaria. Ranatra has also been reported to be an active predator of larvae.

An anthomyid fly (Lispa) has been considered among the insect predators of mosquito larvae but probably this is more important as a predator of adult mosquitos.

Vertebrates. Among the vertebrates, tadpoles may prey on mosquito larvae mainly in small breeding places with very shallow water, but the most important of all predators of mosquito larvae are the mosquito fish (Fig. 8, 9, 10, 11). Gambusia is one of the most efficient species. This is an active top feeder, hunting its prey. This fish has been successfully used in numerous countries in antilarval measures, mainly in those breeding places where there are no natural enemies and where Gambusia have free movement on the surface where anopheline larvae feed.

List of common mosquito predatory fish

Species	Common name
Gambusia affinis (Baird and Girard)	Mosquito fish
Lebistes reticulatus (Peters)	Guppy
Tilapia mosambica (Peters)	Tilapia
Tilapia maciochei (Boulanger)	Tilapia
Tilapia zillii (Gervas)	Tilapia
Tilapia melanopleura (Dumeril)	Tilapia
Cyprinus carpio (Linneus)**	Carp
Carassius juratus (Linneus)**	Goldfish
Xiphophorus maculatus (Gunther)	Moonfish
Nothobranchius guentheri	Annual fish
Cynolebias bellotti	Annual fish
Cynolebias elongatus	Annual fish
Aphanius dispar (Rüppell)	Annual fish
Anabas scandens (Bloch)	Kazari
Aplocheilus panchax (Hamilton)	Indian minnow

** When young; as adults they are not hunters of mosquito larvae.

Much more data exists on the use of Gambusia than on the annual and other fish.

There are about 20 species of annual fish used by aquarists. Cynolebias belongs to an area of South America whereas Nothobranchius originates from Africa (about 12 species have been found in South-East Africa). Both Cynolebias and Nothobranchius reach sexual maturity in about five to six weeks after hatching. Hatching of the eggs will occur even when the eggs are continuously submerged after passing a period of quiescence. The eggs of the annual fish are resistant to dry conditions; when the ponds, ditches and mudholes dry up, the eggs survive in the mud but the adults die. At the beginning of the rainy season, the water fills the seasonal breeding places, the eggs hatch and the young fish grow rapidly.

Fig. 8. Larvivorous fish: <u>Gambusia affinis</u>, left male, right female.

- 27 -

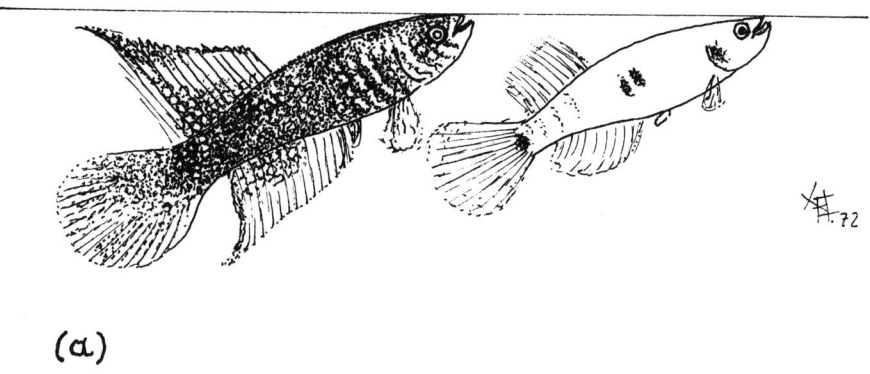

(a)

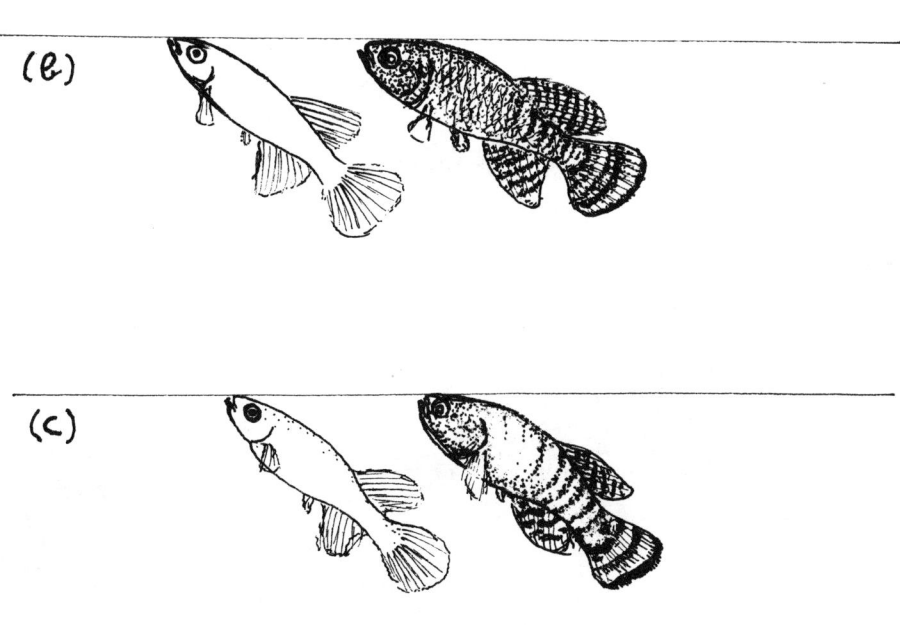

(b)

(c)

Fig. 9. Larvivorous fish: (a) <u>Cynolebias whitei</u>, female and male;
(b) <u>Nothobranchius guentheri</u>, male and female (X 2/3);
(c) <u>Nothobranchius rachovi</u>, male and female (X 2/3).

- 28 -

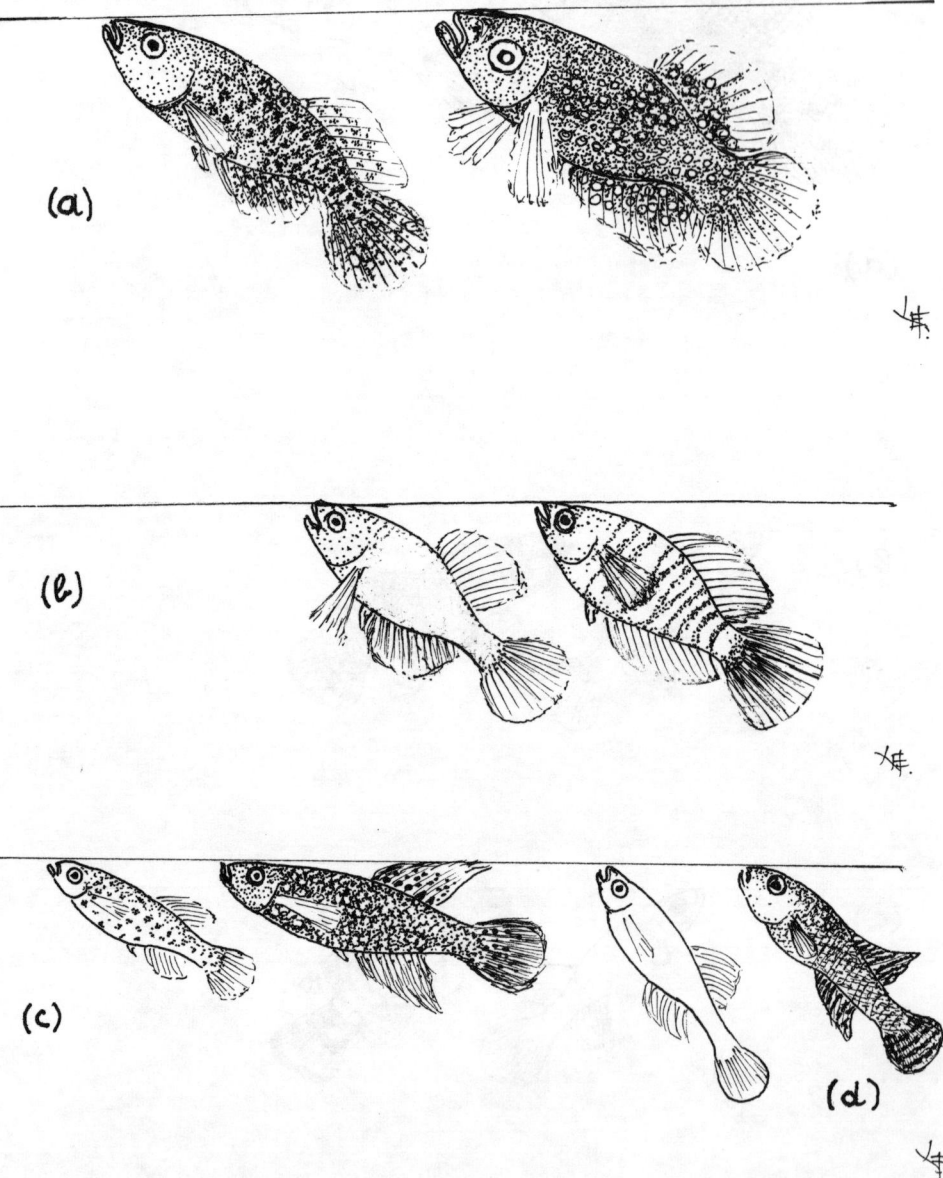

Fig. 10. Larvivorous fish: (a) <u>Cynolebias wolterstorffi</u>, male and female (X 1/2); (b) <u>Cynolebias adloffi</u>, male and female (X 1); (c) <u>Cynolebias melanotaenia</u>, male and female (X 1); (d) <u>Cynolebias ladigesi</u>, male and female (X 1).

- 29 -

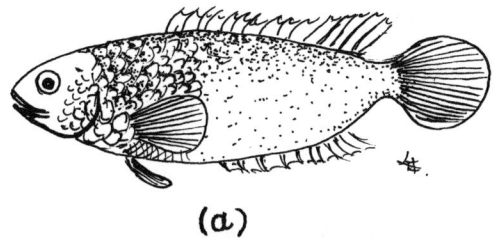

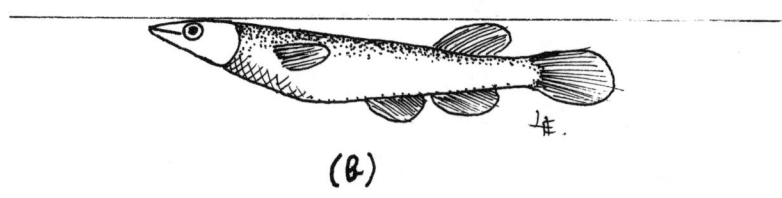

Fig. 11. Larvivorous fish: (a) <u>Anabas scandens</u> (India); (b) <u>Aplocheilus panchax</u> (Java).

5.4 STUDY OF THE LARVIVOROUS ACTIVITY OF VARIOUS SPECIES OF LOCAL FISH

It is well known that many species of local fish eat mosquito larvae during their first stage of life, when they feed at the surface of the water. Many of them do not feed at the surface once they have reached the adult stage. Therefore, a study of predatory habits should be made first on those species which feed all their life at the surface of their habitat. The following steps are suggested for such a study:

Step one. Laboratory observations. Put about 250-500 ml of water in a plastic or glass vessel of about 1000-2000 ml capacity and introduce two to four fish and a known number of larvae. Observe the behaviour of the fish without disturbing them. Count the number of larvae after one, five and 24 hours. If this step indicates that the fish are eating the larvae, proceed to:

Step two. Create a small artificial breeding place, 40-50 cm deep and about 1 m^2 surface area. Add 10 fish and a known number of third and fourth instar larvae of Culex and Anopheles, count the larvae after 24 hours. Direct observation on the behaviour of the fish should also be carried out.

Step three. Select some small natural breeding places with a relatively high larval density and without natural enemies of the fish. Add 5-10 fish per m^2 to some, leaving others as controls. Establish the larval density before adding the fish and 24 hours after and then at 10-15 day intervals. After such a trial, observations should be expanded on the efficacy of fish in various breeding sites representing various biotopes. This step should be carried out during the main mosquito season.

5.4.1 Requirements for an efficient mosquito fish

They must be:

- top feeders, and small, so that they can get about in shallow water;

- able to breed freely in confined waters;

- able to withstand transport and handling;

- difficult to catch, and able to escape their natural enemies;

- worthless or insignificant as food for the human population.

5.4.2 Indications for use

A. As supplementary or main mosquito control measure in:

- reservoirs and other water impoundments. With relatively polluted water, they should be adapted gradually from fresh water to polluted water, and stocks should be created in polluted water before distribution;

- anti-malaria drains under active maintenance;

- irrigation ditches and canals in rice fields, cleared of vegetation;

- wells, cisterns, tanks for domestic water supply, domestic ornamental pools;

- lakes and ponds with clear shorelines;

- streams.

B. **Limitations**

- Fish may have a deleterious effect upon drinking water.

- If the adult fish are cannibalistic and natural food runs short, the young ones are eaten.

- Many species cannot be adapted to brackish or salt water to control mosquito breeding.

- They are only effective if present in large numbers.

- They are only completely effective in the absence of weed and floating debris.

- Small children may catch them if they get the opportunity.

- Periodical inspection is necessary to see that the fish are flourishing. If they are not, the breeding places must be re-stocked.

- Efficiency varies with the season in some types of waters.

In Table 1 a few characteristics of the habitats of the main larvivorous fish are given. Amongst all larvivorous fish, Gambusia is at present employed successfully for the control of mosquitos in some specific situations. It is desirable that the other recognized larvivorous fish should receive increased attention, field trials being necessary to establish their efficacy under various local conditions.

C. **Preparatory steps for the use of Gambusia**

(a) Mapping of the breeding places, their classification and surface. Information on the presence of the natural enemies and larval density.

(b) Evaluation of the approximate number of Gambusia needed:

- for seeding breeding places in order to control the mosquitos immediately.

- for seeding breeding places in order to build up a permanent population of Gambusia gradually.

(c) Creation of a stock population of Gambusia from which the fish will be distributed.

(d) Training of staff and education of the local people.

(e) Various measures to make the breeding places more suitable for Gambusia.

(f) Examine the possibility of transporting Gambusia from the stock rearing places to the areas to be seeded.

(g) Prepare the necessary number of barrels (wooden barrels are preferable) or special containers for the transport of Gambusia, such as double plastic bags placed in boxes, as well as a clean Hudson sprayer to aerate the water in the barrel if the journey lasts more than 2-3 hours. See if ice might be available to reduce the temperature of the water to 20-22°C during transportation on very hot days. In general, the transportation should be done in the early morning or in the evening. Young Gambusia are more resistant than old pregnant females. The addition of oxygen to the plastic bags reduces considerably the mortality of fish during transportation.

Table 1. Main larvivorous fish and their habitats

Species	Climatic zone	Type of water	Type of breeding place	Remarks
<u>Gambusia</u> spp. (top minnow)	Temperate, sub-tropical, tropical	Clear, stagnant or running, fresh water (or with low salinity or low organic pollution)	Permanent ponds, lakes, artificial reservoirs, canals, rice fields, small slow-running streams	In permanent breeding places 200-400 should be introduced per ha initially. When used in temporary breeding places, should be introduced each time in effective numbers, 3-4/m^2. The population is self-maintaining in permanent breeding places only in absence of powerful natural enemies.
<u>Lebistes</u> spp. (Guppy)	Tropical and sub-tropical	Clear or polluted water with low or high salinity	Permanent ponds, lakes, artificial reservoirs, canals, small polluted collections of water (tanks or trenches with polluted water)	As above
<u>Nothobranchius</u> spp. <u>and Cynolebias</u> spp. (instant fish) (annual fish)	Tropical (Africa) Tropical (S. America)	Fresh water, clear or turbid, support moderate salinity	Temporary small breeding places which become dry at certain periods of the year (only the eggs survive this dry period)	Does not always persist in permanent water collections. Is self-maintaining in temporary water collections.

D. Distribution of Gambusia

(a) Timing: In areas with winter, the early spring is best. In sub-tropical and tropical areas, all year round; very early in the morning.

(b) Average number of Gambusia per breeding place:

- for immediate effect two to six female Gambusia per m^2 (in wells, small basins, two per m^2 might be sufficient; in rice fields, pools, five to six Gambusia females per m^2).

- for delayed effect, 200-400 Gambusia, preferably gravid females, per ha. They will build up a satisfactory population in two to three months and then in the absence of natural enemies, they will maintain a high density for many years.

E. Re-seeding of breeding places or reduction of the number of Gambusia when too abundant

- Those breeding places from which the Gambusia have disappeared should be investigated to ascertain the causes, and re-seeded.

- When Gambusia become too numerous, as may happen in wells, small tanks and domestic basins, the number should be reduced by collecting some of them with a net. If too numerous, the adults will eat the young and will become less efficient in hunting the mosquito larvae.

F. Evaluation of the results

(a) Evaluation of the status of the population of Gambusia. Seeded breeding places can be observed at two-week intervals for the presence or absence of Gambusia and their approximate number by counting the "rises" on an approximate area of surface.[4]

(b) Evaluation of the larval density every two weeks during special studies, once a month in other situations.[4]

(c) Evaluation of the relative adult vector density every two weeks and comparing with the data prior to the start of the use of Gambusia, and/or with the data obtained from localities with similar breeding potentialities but without Gambusia.

(d) Parasitological evaluation. This should be carried out by:

- parasitological surveys at the end of the transmission season;

- regular case detection (passive or active or both) as feasible under local conditions.

6. THE INTERACTION OF ENTOMOLOGICAL FACTORS IN MALARIA TRANSMISSION

As explained in the Introduction to Part I, the natural transmission of malaria is influenced by factors related to the parasite, man and his immunity, the vector, and the environment. In the foregoing pages, factors related to the environment have been discussed. The various factors affecting transmission do not act independently. It is the interaction of the factors that shapes the epidemiological significance of a malaria vector in a given area.

---

[4] Once the efficacy of Gambusia under local conditions has been well established, observation on the presence of Gambusia and larval density could be carried out at monthly intervals.

The discussion in this section aims at introducing in colloquial terms the subject of quantitative epidemiology which is thoroughly discussed in Chapter 3.

<u>The entomological factors</u>

- vector density
- frequency of biting man
- longevity

<u>The environmental factors</u>

- temperature
- relative humidity
- rainfall

In order to appreciate the influence of these factors, it is convenient to consider the interaction of some factors:

<u>temperature</u> X <u>frequency of feeding</u> X <u>feeding habit</u> (human blood index)

Assuming that the temperature conditions prevailing in an area allow two days for the gonotrophic cycle to take place, this would mean that feeding of the vector occurs every second day.

If this vector is highly attracted to man, and has the above frequency of feeding, it has a much higher chance of picking up and disseminating the infection than in the case of a vector which has certain zoophilic tendencies.

As temperature decreases, the duration of the gonotrophic cycle increases and the feeding of the vector will be less frequent.

We should now assemble all the entomological and climatological factors, as follows:

Density

X Duration of sporogony <u>versus</u> temperature

X Man-biting habit (frequency of feeding X human blood index) <u>versus</u> temperature and humidity

X Longevity <u>versus</u> temperature and humidity

If optimum temperature conditions prevail in an area over a period of time, completion of the duration of sporogony will be short, for example, about 12 days with <u>P. falciparum</u> and about eight days with <u>P. vivax</u>.

It now depends on whether a proportion of females of the vector population can survive the above-mentioned periods, i.e., until sporozoites appear in the salivary glands, and whether they will survive for a subsequent period. The number of days for which the infective females can survive determines their efficiency in malaria transmission. For example, if the infective females live 10 days <u>after</u> the completion of sporogony, it means that they feed five times during the period, assuming that the gonotrophic cycle remains at two days, thus disseminating the infection in five persons, if they consistently feed on man. This can be compared with another vector having the same density, but which can survive for a period of only two to four days <u>after</u> completion of sporogony. As mentioned above, however, the efficiency of both vectors will be affected if a zoophilic tendency exists because some infective bites will be "wasted" on animals.

Whether or not a proportion of vector population can survive to the completion of sporogony and for a subsequent period of infectivity is generally affected by humidity although species may differ genetically in their adaptiveness to climatic changes. Hot and arid climatic conditions cause high mortality in vector populations. Thus, only a small proportion sheltering in suitable microclimatic niches may be able to complete sporogony. On the other hand, cool conditions would favour survival but extend the duration of sporogony. Obviously transmission cannot take place unless a certain proportion of vector females survive long enough to allow the sporogonic cycle to be completed and infection to be disseminated. This lengthy period needed for sporogony in areas at high altitude or in the spring in temperate zones, which may out-last the life span of the mosquito, accounts for the very low degree or lack of transmission in these areas.

It should be remembered that the vector population does not normally consist of individuals of the same age since the rate of development of immature stages is not even and generations usually overlap, particularly in tropical and sub-tropical conditions. At a given time, under stable breeding and favourable climatic conditions, the vector population should be composed of individuals of all age-groups. Due to natural causes, a certain proportion of the population will die each day, thereby leaving only a proportion of the vector population to survive the duration of sporogony and a subsequent infective period. Depending on the genetic adaptiveness to environmental changes, a vector having a low daily mortality is arbitrarily described as being "long-lived", while a less adapted vector, by being sensitive to changes in relative humidity, suffers from higher daily mortality and is arbitrarily described as being "short-lived".

Obviously, with the high daily mortality there will be few individuals that can maintain transmission. This would need to be compensated by the presence of a high density to effect malaria transmission. However, the probability of the daily survival changes from time to time, depending on climatic conditions and a vector which is particularly "short-lived" may become "long-lived" when climatic conditions favour its longevity.

The two important vectors A. gambiae - in the Ethiopian region - and A. culicifacies - in the Oriental region - are given as examples to illustrate the contrasting characteristics and reaction to the environment, although both are responsible for high endemicity conditions, but of different types, as summarized from Macdonald (1957).

A. gambiae

- high human blood index (about 80%)

- typically long-lived (estimated daily mortality 5-10%)

- sporozoite rate generally high (of the order of 5-10%)

- the critical density required for maintaining transmission is very low (about 0.025 per man per night). Seasonal changes, other than temperature changes, alter the level of transmission, but rarely bring it to a complete end unless the reduction of breeding is very marked.

- connected with stable malaria in equatorial Africa. In these areas, anophelism without malaria has not been recorded, except possibly in off-shore islands, as in Mauritius where malaria has been eradicated.

A. culicifacies

- mainly zoophilic, but may feed predominantly on man when insufficient animals are available as source of food (human blood index 3.3-12.1%)

- typically short-lived (estimated daily mortality 22.5%), but when suitable climatic conditions occur - i.e., when temperature is rather cool - it may be long-lived. Afridi, Majid & Shah (1940) were able to recapture A. culicifacies 40-56 days after liberation.

- sporozoite rate generally low (about 0.1% and as low as 0.064% has been recorded).

- the critical density required for maintaining transmission is much higher than that of A. gambiae (some records give 7-10 and 17-20 per man per hour). The occurrence of seasonal changes in malaria transmission in response to factors other than temperature change is very marked.

- connected with unstable malaria. In such areas, anophelism without malaria is well known. Malaria transmission fluctuates greatly from year to year.

An A. gambiae population can therefore maintain high endemicity in an area, with very limited fluctuation of incidence from year to year. In contrast, A. culicifacies, by virtue of its preference for animals and sensitivity to changes in humidity, can give rise to marked changes in malaria transmission. For example, in the absence of animal hosts it will be forced to feed on man. Higher sporozoite rates were recorded with A. culicifacies especially when climatic conditions favoured its breeding and longevity. Frequent but intermittent rainfall after drought conditions would extend breeding places for A. culicifacies, and hence cause an increase in density. Rain also increases relative humidity, thus favouring higher longevity. In some years, great changes in climatic conditions gave rise to severe outbreaks of malaria.

Examples of other principal vectors that more or less follow the pattern of A. gambiae are: A. funestus, A. minimus minimus, A. fluviatilis, A. sacharovi and A. labranchiae.

Examples of vectors which in general follow the pattern of A. culicifacies are: A. stephensi, A. superpictus, A. pharoensis, A. philippinensis, A. maculatus, A. minimus flavirostris, A. aconitus and A. annularis.

There is still the interaction of the vector population with the other ecological conditions, mainly those which are related to man and his environment, such as housing conditions, economic development including irrigation schemes, and availability of animals, movement of populations, etc. Reference is made here to the role played by an exophilic or partially exophilic vector under poor housing conditions and where humans may spend much of the night outside, with the result of increasing the chance for outdoor transmission of malaria. In addition man-made breeding places would increase vector output and the size of population density. These are both important factors which have to be considered in planning attack measures.

7. VECTOR RESISTANCE TO INSECTICIDES

The genetic basis of resistance to organochlorine insecticides is well established and dependent in most cases on single genes. Evidence exists in A. stephensi that the basic sequence in chromosome arrangement in respect of a particular inversion on chromosome 2R imparts an increased tolerance to insecticides when compared with the inverted sequence. This increased tolerance has been shown not only to DDT but also to malathion. It seems possible, therefore, that the high level of resistance to DDT recorded in some populations of this and other species may be a combined influence of specific genes and a particular chromosome arrangement.

Genetic studies of resistance to organophosphorus and carbamate compounds are only in their infancy. These should be encouraged, particularly to establish cross-resistance patterns with the aim of reducing the number of different insecticides needed for the standard impregnated papers to be used for testing and to reduce the amount of testing.

The various methods of susceptibility testing of adults are as follows:

(a) The exposure to several concentrations for the same length of time and the construction of a regression line from which the LC_{50} is determined.

(b) The exposure to a single concentration for different times and the determination of the LT_{50}.

(c) The determination of the minimum LC_{100} and the establishment of a discriminating dosage near to the minimum LC_{100} in respect of most anopheline species.

The use of the discriminating dosage from the operational point of view has the following advantages:

- The smaller chance of missing the presence of relatively low proportions of resistant individuals; in other words, the earlier detection of resistance.

- The presence of resistant individuals can be demonstrated in small batches of insects: such presence might be missed if exposures were made to serial concentrations.

- A saving of impregnated papers.

- Saving in time in recording and interpretation of data.

7.1 INTERPRETATION OF RESULTS OF SUSCEPTIBILITY TESTS USING ORGANOCHLORINE INSECTICIDES

On the basis of the use of the discriminating dosage, the detection of dieldrin resistance presents very few problems. Within one hour 0.4% dieldrin kills all susceptible genotypes of all anopheline species with the sole possible exception of Anopheles sacharovi, which requires 0.8% for one hour. Survivors of the dieldrin discriminating concentration are almost certainly resistant individuals, i.e., heterozygotes or homozygotes for the resistance gene. If there are any doubts, then testing of the offspring of these survivors on the same discriminating dosage will give no more than 50% mortality (if the survivor is a heterozygote and had mated with a susceptible male - any other mating would produce a lower mortality than this).

With the common incompletely dominant form of dieldrin resistance, a second discrimination between heterozygous and homozygous resistant individuals is possible by exposure to 0.4% dieldrin for two hours (possibly four hours in the case of A. maculipennis). Survivors of this concentration and exposure time will be homozygous resistants, and their offspring should show no mortality on the same concentration and exposure time if the parent female had previously been fertilized by a homozygous resistant male. If fertilized by a susceptible male, the offspring would show 100% mortality on this higher discriminating exposure but no mortality on the lower one (being all heterozygotes). If fertilized by a heterozygous male, then the mortality would still be nil on the lower discriminating exposure and 50% on the higher. The use of other dieldrin concentrations (0.05, 0.1, 0.2 and 1.6%) will give no useful information on the proportion of genotypes present in the population.

The detection of DDT resistance by the use of the discriminating dosage does present some problems because initially the level of resistance is of a much lower order than that shown to dieldrin. A discrete discrimination is not always possible but 4% DDT for one hour represents a useful starting dosage for most anopheline species. In previous years, a purely arbitrary classification of resistance in anopheline mosquitos was based on the percentage of survival to concentrations of insecticides which normally produced a 100% kill. A population was considered susceptible if up to 10% survival was recorded, intermediate if this was between 10 and 50% and resistant if more than 50%. It is now obvious that this classification has little practical value since the intermediate category in particular was not necessarily considered as a case of definite resistance. Instead, the following initial classification is proposed with respect to DDT with the following interpretations:

```
99 - 100% mortality      = susceptible
80 -  98% mortality      = verification required
less than 80% mortality  = resistance strongly suspected but verification still required
                           and field observations to be initiated to determine operational
                           significance.
```

A strong indication of resistance will be given by consistent similar proportions of survivors on exposure to 4% DDT for one, two and four hours as has been generally practised in the field in past years but more certain verification can be carried out by obtaining the offspring of survivors of an exposure to 4% DDT for two hours and rearing these with as little larval mortality as possible by appropriate breeding techniques. Both males and females are then tested, preferably within 24 hours of their emergence by the method adopted by the Ross Institute of Tropical Hygiene, London, on the same concentration and exposure period as was used in the original parent test, namely 4% DDT for two hours. A significantly lower mortality than in the parent population would then confirm the presence of resistance and further selections in the laboratory should produce a fully resistant population. If the test shows mortalities in the offspring similar to those of the parent sample this would be interpreted as a case of tolerance. If this proves impracticable in some situations, it is suggested that the material be sent to special resistance investigation laboratories.

7.2 INTERPRETATION OF RESULTS OF SUSCEPTIBILITY TESTS TO ORGANOPHOSPHORUS (ADULTS AND LARVAE) AND CARBAMATE (ADULTS) INSECTICIDES

A start has been made in the establishment of discriminating dosages for organophosphorus and carbamate compounds using adults of a number of species of anophelines from laboratory colonies at the Ross Institute of Tropical Hygiene, London.

Male and female adults, less than one day old, of the following species of anophelines have been exposed for one hour to 0.01% and 0.1% propoxur, 0.1% and 1.0% fenitrothion and 0.5%, 3.2% and 5.0% malathion at 26°C and 70-80% relative humidity and held in these conditions for a further 24 hours, when mortalities were determined - all observations were made between March and August 1973:

A. gambiae species A	- DDT and dieldrin-resistant from Togo
A. gambiae species B	- DDT and dieldrin-resistant from Sudan
A. gambiae species C	- from Southern Rhodesia and Transvaal, South Africa
A. melas	- from The Gambia
A. merus	- from Tanzania
A. stephensi	- 1947 Delhi strain ex-Horton Hospital, Surrey, England
A. stephensi, 2RA	- DDT and dieldrin-resistant from Mamlaha, Iraq, with the basic chromosome 2 arrangement
A. stephensi, 2RB	- DDT and dieldrin-resistant from Mamlaha, Iraq, with an inverted chromosome 2
A. farauti No. 1	- from Rabaul, Papua New Guinea
A. farauti No. 2	- from Queensland, Australia
A. balabacensis	- from West Perlis, Malaysia
A.l. atroparvus	- English strain, ex-Horton Hospital, Surrey, England
A.l. labranchiae	- from Algeria
A. sacharovi	- from Turkey
A. quadrimaculatus	- DDT and dieldrin-resistant from the United States of America
A. albimanus	- from Haiti and Panama

With few exceptions, complete kills or only single survivors were obtained with 0.1% propoxur, 1.0% fenitrothion, and 5.0% malathion.

The exceptions were:

A. sacharovi on 0.1% propoxur - only 61% mortality after 24 hours and 79% after 48 hours. The surviving mosquitos were obviously affected and had their wings permanently splayed.

A. balabacensis on 1.0% fenitrothion - only 73% mortality.

A. quadrimaculatus on 5.0% malathion - only 87% mortality.

3.2% malathion failed to kill all A. stephensi 2RA (only 39%), A. quadrimaculatus (79%), A. balabacensis (89%), A. sacharovi (92%) and A. merus (96%).

0.01% propoxur, 0.1% fenitrothion and 0.5% malathion usually give quite low mortalities of anopheline mosquitos.

Laboratory indications are that only 30 minutes exposure to 0.1% propoxur is necessary to kill susceptible Anopheles albimanus but some investigators using the same species in the field are of the opinion that longer exposures even up to four hours may be necessary and that variations in temperature and humidity can significantly affect results. However, longer exposures applied in general to other species may result in masking the presence of resistant individuals. Therefore, until further experience is gained, 0.1% propoxur for one hour is recommended.

Discriminating dosages of organophosphorus compounds remain to be established in respect of larval testing. This is especially necessary for Abate which is used exclusively as a larvicide in several malaria programmes.

Larval tests should be conducted when larvicides are being used while adult testing is appropriate when the insecticide is used as an imagocide. For rapid detection of resistance, it is preferable to obtain mosquitos which have already been in contact with the insecticide, i.e., from sprayed houses or from treated breeding places, in other words which consist of an already selected sample. To determine the true proportions of resistant and susceptible individuals in a population, the best method for obtaining unbiased samples is to breed out adult mosquitos from larvae or collect from baits.

The influence of temperature and humidity on the results of susceptible tests is very important. DDT is found to be less toxic at high temperatures while the reverse is the case with organophosphorus compounds. There are some indications that testing with carbamate insecticides is influenced by temperature and humidity as well. Recent observations showed that the susceptibility level of A. aquasalis was found to vary according to the time of day the mosquitos were exposed to DDT. Mosquitos exposed in the morning hours were less susceptible than those exposed during the evening hours. The influence of circadian rhythms of mosquito activity in susceptibility testing should be explored in more species and the time of day of testing should be standardized, if possible, in order to obtain comparable results.

Difficulties in the detection of DDT resistance (and possibly also resistance to organophosphorus compounds and carbamates) are acknowledged. Since the testing of survivors of discriminating dosages presents some problems, perhaps the answer will be to have special insecticide resistance investigation laboratories, where questions of resistance unresolved by the methods already advocated might be dealt with. Such a laboratory would have to be equipped for rearing and colonization (including facilities for artificial mating) and in them prolonged selection of suspected populations could be carried out.

Due consideration should be given in the siting of these reference laboratories so that minimum delays occur in posting egg samples. Mortalities in transit could differentially affect survival of resistant phenotypes.

7.3 DETERMINATION OF THE OPERATIONAL AND EPIDEMIOLOGICAL IMPLICATIONS OF INSECTICIDE RESISTANCE

It need hardly be said that the appearance of dieldrin resistance necessitates a rapid change of insecticide and that no other cyclodiene insecticide or HCH can be used as an alternative. The first alternative must be DDT. In fact, this is practically always the first choice in new spraying campaigns. The appearance of DDT resistance, however, need not necessitate the immediate abandonment of this insecticide. This resistance when it first appears is usually of such a low degree that a proportion of resistant individuals may still be killed. With the less efficient vectors, the mortality, albeit reduced, may still be sufficient to intercept malaria transmission. What is required when this situation develops - and this may also apply to resistance to organophosphorus compounds and carbamates where so far it appears that levels of adult resistance are low - is some system of monitoring the actual mortality occurring under field conditions to see that the mosquito population does not reach an epidemiologically significant level of density and longevity. It may well be that the interruption of transmission is maintained long enough for parasite eradication to be achieved. On the other hand, a situation like the one in the Persian Gulf area in 1966 may be reached and a change to a group of insecticides other than the organochlorines or to some other control method will have to be made.

It is suggested that the following procedures be intensified in a selected study area as soon as the presence of the type of low-level resistance described is indicated (more than 20% survival on 4% DDT for one hour, for example):

(a) It is of particular importance to ensure that total coverage spraying in the selected area be well maintained. This will eliminate the operational deficiencies, hence the implications of insecticide resistance alone on malaria transmission can be clearly demonstrated.

(b) Indicator villages should be selected in areas representing the main ecological conditions of the sprayed area and possibly different levels of resistance.

(c) A number of houses (usually 10) in each village should be selected for the installation of window traps and observations on species and numbers of mosquitos caught in these traps and their subsequent mortality made at regular intervals (twice a week). Ideally, floor counts of dead mosquitos should be made but this would normally require the construction of special ant-proof trap huts. In their absence, window trap delayed mortality should be taken as an indicator of kill.

(d) A search should be made for daytime house-resting mosquitos when the vector is mainly endophilic; their presence is a strong indication of resistance where DDT is the insecticide being used. DDT, it will be remembered, is normally highly irritating to susceptible mosquitos and daytime house-resting mosquitos are virtually never seen in DDT-sprayed houses in the absence of resistance.

(e) Female mosquitos caught in houses or in window traps should be classified according to their stomach and ovary status. Trap observations are most appropriate in the case of exophilic species.

(f) Man-bait collections should be made periodically both indoors and outdoors and their parous rate determined.

(g) Susceptibility tests using discriminating concentrations should be carried out on a widespread sample of vectors at the beginning and towards the end of the spraying cycle.

(h) In any eradication programme, careful epidemiological investigation of malaria cases should be carried out, paying particular attention to determining the date of onset of the primary attack and its relation to the date of spraying and the density and longevity of the vector. In areas under early attack, periodic blood examinations should be carried out to monitor changes in parasite rates.

Resistance levels determined by larval susceptibility tests may not be relevant in implications of resistance in adults, e.g., resistance levels in larvae of Aedes aegypti to DDT and Anopheles albimanus to OP compounds and carbamates would seem to be much higher than in adults of these species.

The decision to change the insecticide must never be taken lightly in view of the very high cost of alternatives, e.g., malathion being five times, and propoxur twenty times as costly to apply as DDT, and in view of the limited number of suitable alternatives available.

Of major concern nowadays is the number of cases of insecticide resistance in malaria vectors directly attributable to the massive use of large numbers of different insecticides for agricultural purposes. The multi-resistance state of A. albimanus in Central America is the latest case in point. More knowledge is needed of cross-resistance patterns in respect of these resistances.

Useful information on the operational life of new insecticides might be gained from laboratory selection of the species against which the insecticide is intended to be used, in order to assess the resistance potential. However, such selections should follow initial field selections of large numbers of wild-caught females exposed to high selection concentrations. However, it should be recognized that what can be predicted by such special procedures may not be applicable to selection occurring under field conditions.

Investigations have been started on the potential effectiveness of presently used insecticides in ultra-low-volume (ULV) applications. Some of these insecticides applied in this way may produce higher kills of resistant mosquitos than conventional formulations. ULV applications are also being investigated as a means of house spraying. Susceptibility testing in respect of this method of application may have to be of a different type from that normally employed.

8. CYTOGENETICS IN STUDIES OF MALARIA VECTORS[5,6]

8.1 INTRODUCTION

The principal aim of this section is to provide field entomologists with the necessary background knowledge in cytogenetics, and the respective methods used, in order that they can follow up the recent developments in cytogenetic studies as applied to malaria vectors. Also as field entomologists are in the best position to detect diversified behaviour patterns in their local vectors, this knowledge will also enable them to contribute effectively to the current studies and may lead to discovery of a group of sibling species or chromosomal types within one species.

---

[5] Extracted from "Review of Cytogenetic Studies on Anopheline Vectors of Malaria" by White, G. B., Coluzzi, M. & Zahar, A. R. (to be issued in the WHO/MAL series of documents in 1975).

[6] For a description of techniques and procedures see Part II, section 13. For further reading see the following references at the end of the Chapter - White (1961, 1973); Srb et al. (1965); Swanson et al. (1967); Wright & Pal (1967); and Wld Hlth Org. techn. Rep. Ser., No. 398, 1968.

The pioneer work by Frizzi in Italy (1947, 1949) on the Anopheles maculipennis complex and the contributions of Kitzmiller and his collaborators on the Neartic members of the same group demonstrated the value of the cytogenetic approach in identifying the species complex and in studying the evolutionary problems by cytotaxonomic methods.

By studying the banding pattern in the chromosomes of the salivary glands of the natural population it was possible to identify the members of the above-mentioned sibling species. The relationship between the Paleartic and Neartic groups has been discussed by Kitzmiller et al. (1967) and a review of mosquito cytogenetic studies has been made by Kitzmiller (1967).

The five sibling species of the Anopheles gambiae complex, the major vectors in Africa, were separated by crossing experiments, Davidson & Jackson (1962), Paterson et al. (1963) and Davidson (1964). Extensive morphological studies by Coluzzi (1964) yielded useful biometral characters for identifying the saltwater species of the complex of West and East Africa, Anopheles melas and Anopheles merus, respectively.

Further attempts to identify the freshwater species were not successful except for also differentiating the saltwater species from the freshwater species (Ismail et al., 1968) or were of limited or no value in differentiating the freshwater species (Chauvet et al., 1969 ; Zahar et al., 1970 ; Clarke, 1971 ; White & Muniss, 1972).

The cytotaxonomic method of studying the banding pattern of the chromosomes of the larval salivary glands as adopted by Coluzzi (1966) and Coluzzi & Sabatini (1967, 1968) and Coluzzi et al. (1969) has proved its practical value in identifying the five sibling species of this complex. Recently the identification has been extended to the adult female by using the polytene chromosomes of the ovarian nurse cells, Coluzzi (1968). Consequently, more identifications could be done adding more knowledge on the geographical distribution of the sibling species of the A. gambiae complex. Besides which spot surveys and longitudinal observations were conducted, though on a limited scale, but have thrown the first light on the seasonal distribution, feeding behaviour and infectivity of certain members of the complex, namely species A and species B, (Coz, 1973a, 1973b; Service, 1970a, 1970b, 1972; White et al. 1972a, 1972b; Shidrawi, 1972; White & Rosen 1973). Further, a new member of the A. gambiae complex, designated species D by Davidson & White (1972) and Davidson & Hunt (1973), was cytotaxonomically studied in relation to species B and C by Hunt (1972).

Cytogenetical studies were extended to other sibling species such as the Anopheles punctulatus complex and two species have been identified, namely the New Guinea, Anopheles farauti No. 1 and the Australian A. farauti No. 2, confirming the crossing experiments that the two are not conspecific (Bryan, 1970; Bryan & Coluzzi, 1971). More recently, cytogenetic methods were employed in the study of the relationship of different chromosomal types in relation to behaviour and morphological characters by Coluzzi (1972), Coluzzi et al. (1972), and Coluzzi (1973). Also populations of A. darlingi originating from different areas in Brazil were examined by Kreutzer et al. (1972). Further, a study has been carried out on A. nuneztovari from Brazil, Venezuela and Colombia (Kitzmiller et al. 1973).

8.2 THE POLYTENE CHROMOSOMES

It is necessary to explain first the structure of the polytene chromosomes in order to understand how the identification of the sibling species could be made. Polytene chromosomes are also called giant chromosomes as they may be 200-300 times bigger than mitotic chromosomes. The banded polytene chromosomes are found in some tissues of insects of the Order Diptera, such as the rectum, Malpighian tubules, larval salivary glands and adult ovarian nurse cells.

The correspondence between the normal mitotic chromosomal complement and the polytene complement is diagramatically shown in Fig. 12. All Anopheles species so far studied show six chromosomes in the mitotic karyotype, i.e., three pairs of homologous chromosomes. Each chromosome has a centromere appearing as a constriction and dividing the chromosome into two arms. Generally speaking, the position of the centromere (more or less central of terminal)

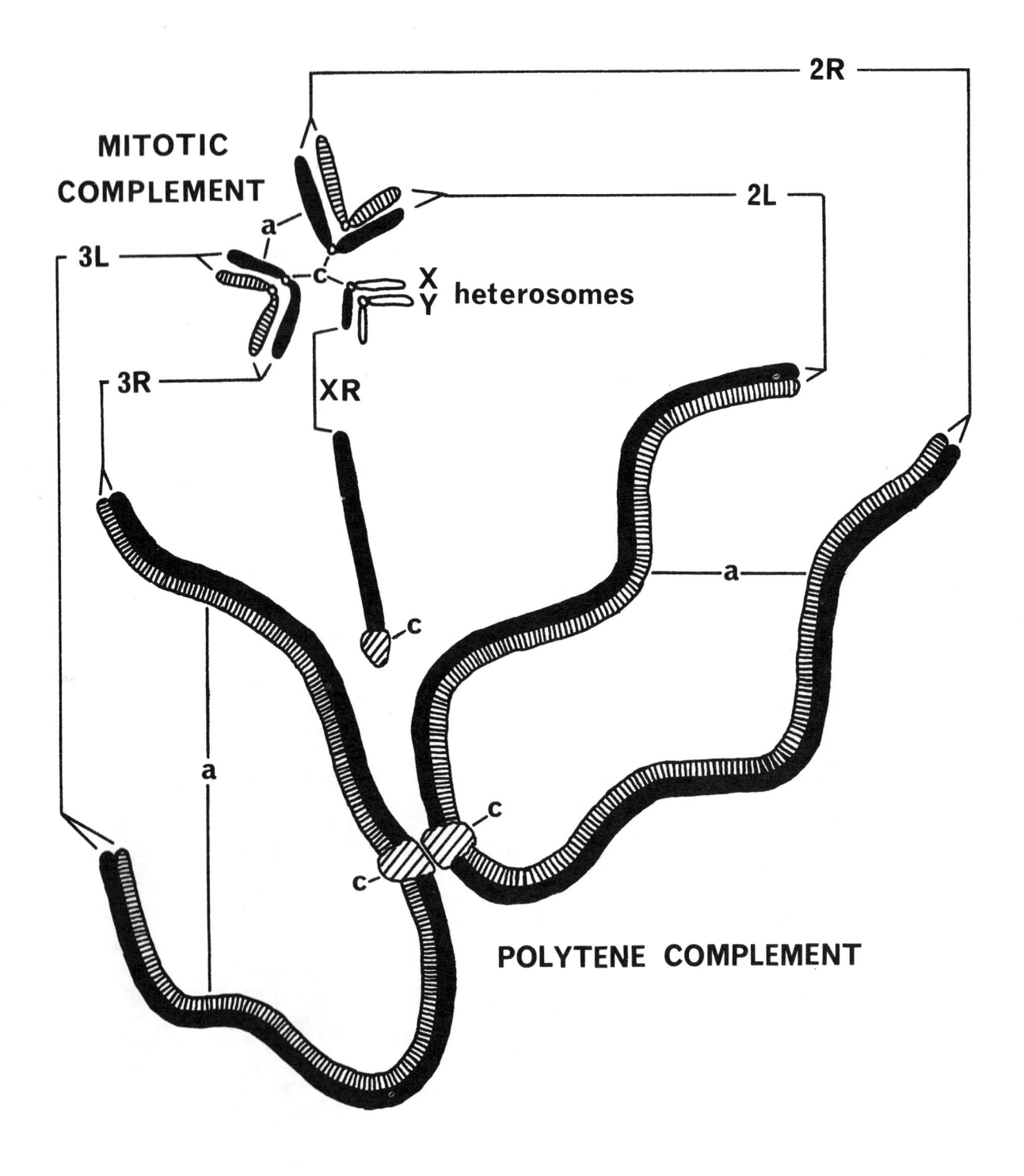

Fig. 12. Correspondence between the three chromosome pairs in the mitotic complement of a male Anopheles and five arms seen in the polytene complement. (a = autosomes; c = centromeres). Autosomal arms are euchromatic in both complements, while the Y-chromosome and the left arm of the X-chromosome are heterochromatic (white segments) and are not represented by a band sequence in the polytene complement. The mitotic and polytene complements are not represented according to their relative size.

determines the shape of the chromosome. Thus, depending on the position of the centromere, the chromosome may vary in shape and hence be called telocentric when the centromere is terminal, or subtelocentric when the centromere is subterminal, or metacentric when the centromere is at or near the middle, or submetacentric. In the mitotic complement of Anopheles, it is generally possible to recognize two autosomal pairs (chromosome 2 and 3) and one pair of heterosomes (chromosome 1 = X and Y sexual chromosomes) often subtelocentric. The two pairs of autosomes are nearly equal in length and can be differentiated as one is more or less metacentric and the other submetacentric. The X chromosome has generally one arm heterochromatic while the Y chromosome is completely heterochromatic. The other arm of the X chromosome and the autosomes are euchromatic being the only recognizable parts of the polytene complement. The polytene chromosomes remain synapsed and each chromosome consists of spiralized DNA helical filaments known as chromonemata (singular = chromonema).

There is now conclusive evidence that the polytene chromosomes originate from a process of reduplication of the chromonemata without corresponding cell division (endomitosis). The homologous chromonemata remain in constant synapsis and produce a polytenic (i.e., multistrand) structure along which a series of dark stained bands alternates with clear interband zones.

Crick (1971) suggested that the chromosomal DNA falls into two classes, termed for convenience "globular control DNA" occurring in the bands, and a much smaller fraction "fibrous coding DNA" occurring in the interbands. The "globular control DNA" of the bands could regulate the activity of the coding DNA in the interbands.

The three chromosomal pairs in the mitotic complement of Anopheles appear as five arms in the polytene complement which are named as follows:

Heterosomes: Chromosome pair No. 1 = X (right and left arms can be detected in only a few species)

Autosomes: Chromosome pair No. 2 : Right arm (2R) and Left arm (2L)
              Chromosome pair No. 3 : Right arm (3R) and Left arm (3L)

The naming of right and left chromosome arms is the prerogative of the person who first describes them.

In squashed preparations, chromosome X usually appears isolated while the autosomal arms often remain attached to the centromere as in Fig. 12.

The banding pattern of the polytene chromosomes is constant corresponding to the gene sequence. Thus, any change in gene sequence might be detected by direct comparison of chromosomes from different individuals or by study of synapsis and band correspondence in the hybrids.

Crossing experiments take a long time and much effort. Attempts to find reliable morphological features for identifying the sibling species of the A. gambiae complex have so far been unsuccessful. The work of Coluzzi & Sabatini (1967, 1968) made it possible to distinguish between individuals of the freshwater species A, B and C through the polytene chromosome banding pattern of chromosome X (Fig. 13). The method for making the nurse cell chromosome preparations is easy to learn and is fully described below.

However, the study of polytene chromosomes, despite its importance, may not provide a reliable distinction of all sibling species. Carson et al. (1967) found known groups of the species Drosophila without detectable change in the banding sequences of any of the polytene chromosomes. These he called homosequential species. An example of homosequential species in anophelines is the case of Anopheles labranchiae and Anopheles atroparvus in which the salivary gland chromosomes are almost identical, with possibly minor rearrangements (Frizzi, 1949). The close correspondence of the banding pattern in these two species was confirmed

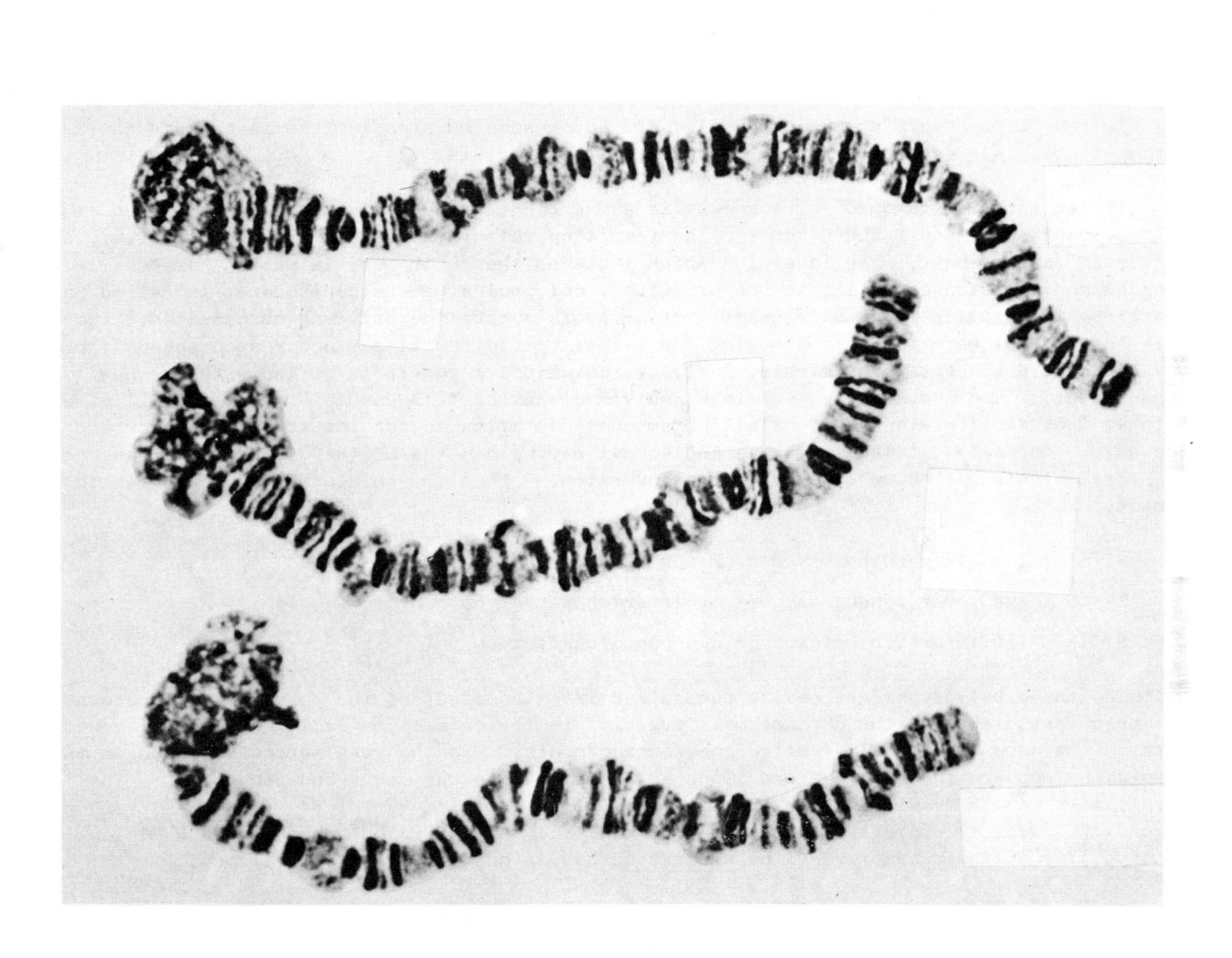

Fig. 13. A. gambiae complex. Ovarian polytene X-chromosomes of A. merus (bottom), species B (top) and A. melas (middle). The X-chromosome of A. merus has the same band sequence as that of species A, while the X-chromosome of A. melas shows the same banding pattern found in species C and D.

by Coluzzi & Coluzzi (1970) who found that the chromosomal arms appeared generally in intimate pairing in the F_1 hybrids and a constant area of asynapsis was observed only in the basal part of X chromosome. Although this aberration indicates genetic differences, the differences in the banding pattern seem too slight to be reliably used as taxonomic characters. Therefore resort to crossing experiments will have to be made and biochemical means may provide a satisfactory answer.

8.3 CHROMOSOME INVERSIONS

As mentioned earlier, the study of the banding pattern of the polytene chromosomes permits detection of chromosomal mutations such as inversions, translocations, deletions, etc. (for definitions see, WHO Technical Report Series, No. 398, 1968). Such aberrations may occur spontaneously or may be induced by certain chemicals, heat, or irradiation.

An inversion occurs when a section of the chromosome rotates 180° compared with the normal arrangements. An example is shown schematically in Fig. 14.

If the normal "standard" arrangement is A B C D E F G and the chromosome is broken at two points - between A and B and E and F - the resulting pattern will be A E D C B F G, i.e., an "inverted" arrangement. An inversion which includes the centromere is called pericentric, whereas an inversion occurring in one arm, i.e., not including the centromere, is called paracentric. The latter type is frequently observed in anopheline polytene chromosomes. When this chromosomal mutation arises during the meiotic division (it may occur spontaneously or may be induced by certain chemicals, heat, or radiation) a gamete is formed with the new arrangement. The probability that this individual having this gamete "inverted" will mate with an individual having an identical chromosomal mutation is too low to be taken into account. Normally, it mates with an individual having gametes of the "standard" arrangement. The first zygote will have a heterozygote inversion. Thus the population may comprise three karyotypes:

ST/ST : the standard homozygote or homokaryotype

IN/IN : the inversion homozygote or inverted karyotype

ST/IN : the inversion heterozygote or heterokaryotype

The three karyotypes are easily recognized from the study of the polytene chromosomes. The principal effect of the chromosomal inversion is to restrict or avoid exchange of genetic material between the two alternative gene arrangements. So the gene series included in the inversion loop are inherited as one block and gene changes may occur independently. The final result will be the coexistence in the same population of three karyotypes (the standard homozygote = ST/ST; the inverted homozygote = IN/IN, and the inversion heterozygote = ST/IN) which differ in their adaptive value vis à vis the various ecological situations.

Inversions are a form of chromosomal mutation that may arise in gametes spontaneously or in response to stresses such as heat, radiation or mutagenic chemicals. A new inversion may soon become stabilized at some frequency in a population if heterozygous individuals have some advantage of fitness over homozygous standard or homozygous inverted individuals. If the homozygous inverted arrangement is superior, it may quickly replace the original standard arrangement completely by means of natural selection. Viable inversions therefore constitute important evolutionary steps. Species having a very marked occurrence of intraspecific inversion polymorphisms (e.g., Anopheles punctipennis, Baker & Kitzmiller, 1964; Anopheles darlingi, Kreutzer et al. 1972; Anopheles nuneztovari, Kitzmiller, 1973 and species B of the A. gambiae complex, Coluzzi & Sabatini, 1967) are considered to be endowed with strong adaptability (Dobzhansky, 1951; Maynard Smith, 1966).

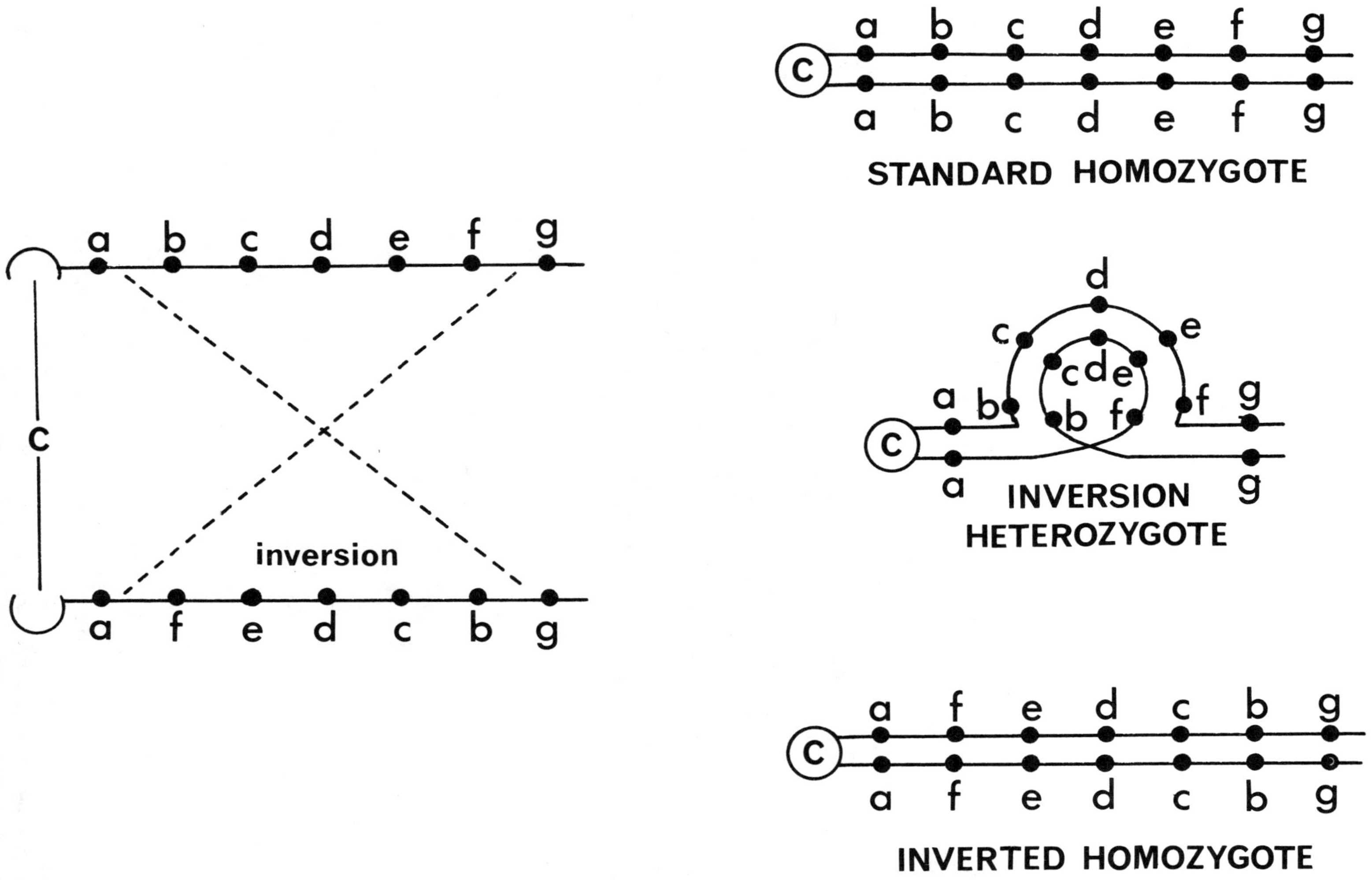

Fig. 14. Diagram of chromosomal rearrangements in inversions.

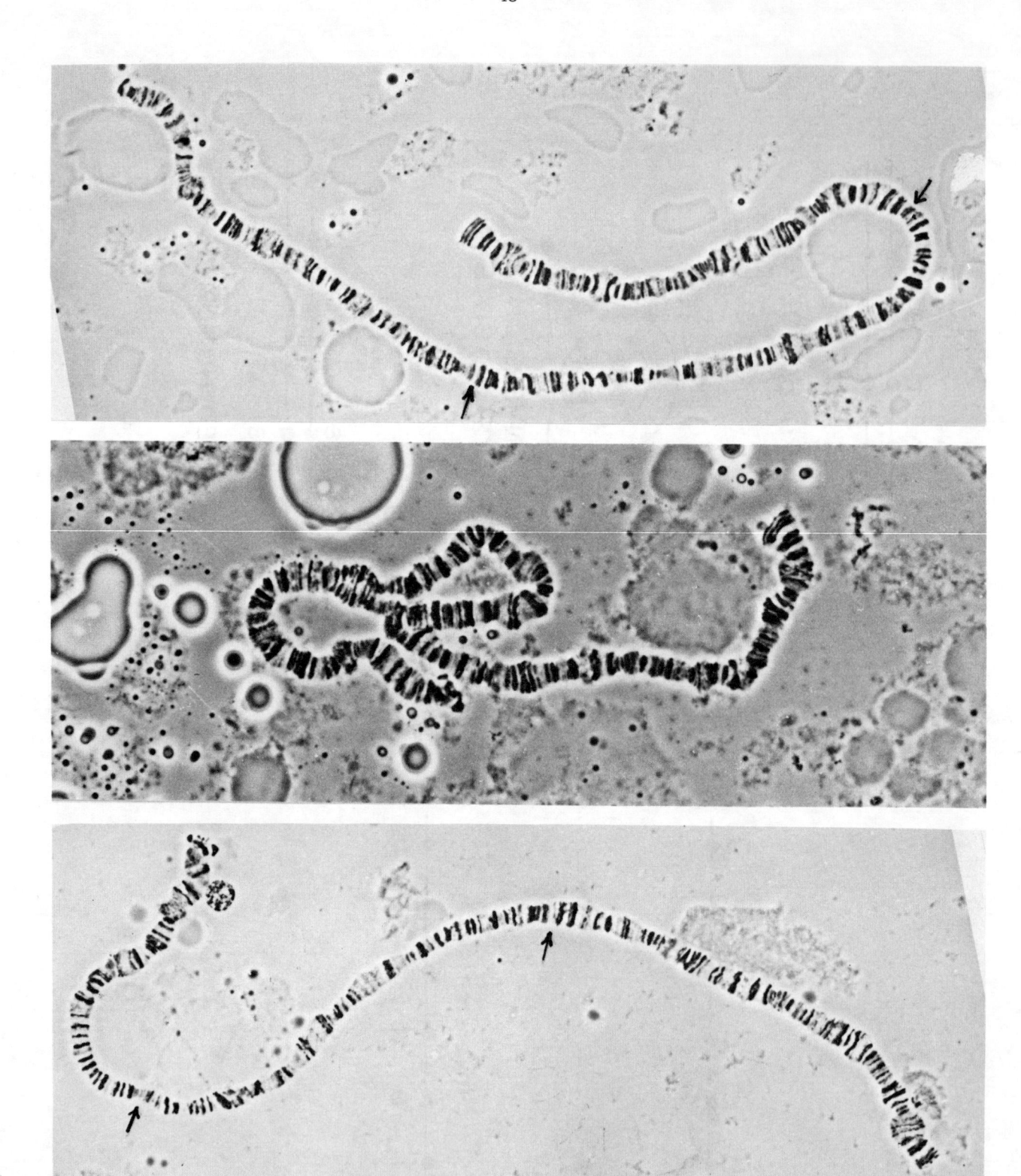

Fig. 15. Inversion "b" in the chromosome arm 2R of <u>Anopheles stephensi</u>. The middle picture shows the inversion loop of the karyotype b/+. The two alternative karyotypes +/+ and b/b are shown above and below respectively, with break points of the inversion indicated by an arrow.

Inversion nomenclature as utilized for Drosophila by Wasserman (1963) and Carson et al. (1967) has now been adopted for anophelines (e.g., Anopheles stephensi and Anopheles superpictus, Coluzzi et al. 1973 and the A. gambiae complex, Coluzzi et al. unpublished data). The system allows a common nomenclature for pursuing interspecific and intraspecific inversions in a group of related species. The polytene sequence of one of the species of the group is chosen as standard. Each new sequence found in the same or another species of the group is described in terms of those segment(s) of the standard sequence which would need to be inverted to produce the observed arrangement. Each inversion is designated by a lower case letter following the chromosome in which it occurs. Polymorphic inversions are indicated by the heterozygous symbol a/+, with "+" used to denote the uninverted or basic sequence. (For further details of inversion nomenclatures, see Coluzzi et al. 1973.)

8.4 RECENT DEVELOPMENT OF THE CYTOGENETICAL STUDIES IN ANOPHELINE VECTORS OF MALARIA WITH SPECIAL REFERENCE TO INVERSION POLYMORPHISM

Apart from the interest in evolutionary relationship between chromosomal rearrangements and speciation, the understanding of the bearing of inversion polymorphism on the biology and behaviour of malaria vectors may unravel facts that would lead to better understanding of the epidemiology of malaria transmission. Although the subject is in its preliminary stages, it is hoped that with increasing interest of workers in this field and intensified efforts, progress will be made in a shorter time than would otherwise be anticipated.

Laboratory and field investigations showed that important biological and morphological divergencies are exhibited by carriers of different chromosome arrangements, (Coluzzi, 1972, 1973; Coluzzi et al. 1972, 1973 and unpublished data) as summarized below.

Laboratory observations. Experiments designed to reveal relationships between inversion arrangements and characters of morphology, behaviour and circadian rhythm were conducted on A. stephensi. The material tested was a strain from Iraq carrying a polymorphic inversion designated "b" (Fig. 15). Pure lines for the b and + arrangements had been selected from the same original strain.[7] The following characters have been shown to be as influenced by these arrangements of chromosome 2R.

(a) Egg size. Female carriers of the two alternative homozygous arrangements produce eggs of significantly different mean length, while the heterozygous females produce eggs of intermediate length. The inversion polymorphism therefore could be related to the two morphological forms of A. stephensi which have been described as mysorensis and the type form.

(b) Mating behaviour. The propensity to mate with males of another species (A. merus) is significantly higher in 2Rb homozygous females than in the 2R+ homozygous ones. The heterozygous females were intermediate in their mating versatility.

(c) Periodicity. Mosquitos homozygous for the inverted or standard chromosome arrangement were found to have distinct daily peaks of pupation and emergence, the heterokaryotype being intermediate between the two homokaryotypes. Differences in flight activity between the two homokaryotypes were also recorded (Jones & Gubbins, In press).

Field observations. In 1971 and 1972, data were obtained from the WHO Malaria Control Project (IR-0172) area in Garki District, near Kano, Nigeria, where both species A and B of the A. gambiae complex are seasonally abundant (Coluzzi, 1973 and unpublished data). Species B is particularly rich in chromosomal inversion polymorphism, with a total of twelve different intraspecific inversions present of which seven involve sections of the chromosome arm 2R.

---

[7] The arrangements of chromosome 2R termed STANDARD and KARACHI by Coluzzi (1972) have been re-named 2Rb and 2R+ respectively by Coluzzi et al. (1973).

Frequencies of the various inversion arrangements are being compared in samples of female mosquitos collected in different resting sites, biting different hosts, biting indoors and outdoors, at different times of night and so on. Results of this long-term study are very complicated and not yet complete, but already show some significant contrasts in the chromosome inversion arrangements between samples positive for human and animal blood, biting at different times of night and resting indoors versus outdoors. To give one example, females carrying the inversion designated 2Ra in the homozygous standard state and/or inversion 2Rc in the heterozygous state are more exophagic and exophilic than the carriers of other 2R chromosomal types.

Recent studies on inversion polymorphism in two malaria vectors in South America mark another important step towards unravelling the reasons for the varied vectorial patterns displayed by different populations.

The work of Kreutzer et al. (1972), on material of A. darlingi from different areas in Brazil, has shown that the northern population (Amazonas) are more polymorphic than the southern populations and it is anticipated that the high amount of chromosomal polymorphism might possibly be linked with vectorial capacity. A further study on A. nuneztovari showed that X chromosome differences between Brazilian and Venezuelan/Colombian populations provide the only reliable means at present for separating vectors from non-vectors. Mosquitos from both areas are indistinguishable morphologically and their behavioural characteristics showed considerable variations.

8.5 FUTURE RESEARCH TRENDS

Field entomologists are well situated to detect variabilities in vector populations that may lead to the discovery of previously unrecognized sibling species or chromosome variants.

For further development of cytogenetic studies in malaria vectors, the requirements are:

(i) Experimental studies under laboratory conditions

Laboratory breeding of strains originating from different geographical areas would make available material that is genetically controlled. This would enable comparison between two experimental populations maintained in habitats which can be altered in a certain definite way.

(ii) Field investigations

Field work is essential in order to collect samples of vector populations for cytogenetical observations on successive generations from different biotopes.

In these samples, the banding pattern of the chromosomes should be studied with a view to recording chromosomal inversion occurring inter- and intraspecifically.

Such a study may lead to the discovery of sibling species which can be confirmed by crossing experiments. Also, it may elucidate patterns of behaviour in relation to inversion polymorphism, as in the case of the field investigations on species A and B of the A. gambiae complex which have been pursued in Garki District, Kano State, Nigeria, with the following aims:

- with a baseline being established cytogenetically, to investigate what the response of species A and B would be to insecticide treatment.

- to determine whether carriers of certain chromosomal arrangements show preferences for human or animal hosts, for resting indoors or outdoors, or for a certain type of breeding place.

- to determine the vectorial efficiency of different chromosome types in malaria transmission.

- to determine whether all the chromosome types will react in the same way to the insecticide treatment.

8.6 PARTICIPATION OF FIELD ENTOMOLOGISTS IN CYTOGENETIC STUDIES

As mentioned above the entomologists in the field observe various behavioural patterns exhibited by local vectors. When abnormalities occur either in natural populations such as changes in the feeding and resting behaviour representing refractory response to insecticidal attack, it would be advantageous if such populations could be examined cytogenetically. For this to be achieved, the following essential requirements must be met:

(a) A standard chromosomal map for the _Anopheles_ vector concerned, or at least one member of the sibling species complex should be available. Standard maps are readily available for a number of malaria vectors, but chromosomal maps are awaited for some vector species, such as _Anopheles annularis_, _Anopheles balabacensis_, _Anopheles culicifacies_, _Anopheles fluviatilis_, _Anopheles funestus_, _Anopheles pharoensis_, and _Anopheles philippinensis_.

(b) The banding patterns that can be used to distinguish between sibling species must be outlined.

(c) Entomologists must be trained to make chromosomal preparations, to read chromosome maps and to locate the banding patterns which can be used to distinguish the sibling species.

It may not be possible for all entomologists to be trained in reading the chromosome maps and in identifying members of a sibling species by the banding patterns. With this in mind, Part II of this Manual gives a detailed description of the technique for making nurse cell chromosome preparations (Part II, section 13).

Using this technique field entomologists may investigate vector populations showing certain behavioural abnormalities and send the preparations to specialists to be checked. Maintaining liaison between the field and laboratory research workers on cytogenetics will open the scope for studies to be extended to many vector species in different geographical areas. In sampling for cytogenetic studies it is essential to record the biotope from which mosquitos were collected, also the prevailing meteorological and other ecological conditions.

BIBLIOGRAPHY

Afridi, M. K. et al. (1940) Studies on the behaviour of adult Anopheles culicifacies, Part II, J. Malar. Inst. India, 3, 23

Baker, R. H. & Kitzmiller, J. B. (1964) Salivary gland chromosomes of Anopheles punctipennis, J. Hered., 55, 9

Bryan, J. H. (1970) A new species of the Anopheles punctulatus complex, Trans. roy. Soc. trop. Med. Hyg., 64, 28

Bryan, J. H. & Coluzzi, M. (1971) Cytogenetic observations on Anopheles farauti Laveran, Bull. Wld Hlth Org., 45, 266

Carson, H. L. et al. (1967) Karyotypic stability and speciation in Hawaiian Drosophila, Proc. nat. Acad. Sci., 57, 1280

Chauvet, G. et al. (1969) Validité d'une méthode chétotaxique de distinction des larves des espéces A et B du complexe Anopheles gambiae Giles à Madagascar, Cah. ORSTOM, ser. Ent. med. et Parasitol., 7, 51

Clarke, J. L. (1971) Potential use of the spermatheca in the separation of species A and B females of the Anopheles gambiae complex in Northern Nigeria, Bull. Wld Hlth Org., 45, 260

Coluzzi, M. (1964) Morphological divergences in the Anopheles gambiae complex, Riv. Malar., 43, 197

Coluzzi, M. (1966) Osservazioni comparative sul chromosoma X nella specia A e B del complesso Anopheles gambiae, Rend. Accad. Naz Lincei, 40, 671

Coluzzi, M. & Sabatini, A. (1967) Cytogenetic observations on species A and B of the Anopheles gambiae complex, Parassitologia, 9, 73

Coluzzi, M. (1968) Cromosomi politenici delle cellule nutrici ovariche nel complesso gambiae del genere Anopheles, Parassitologia, 10, 179

Coluzzi, M. & Sabatini, A. (1968) Cytogenetic observations on species C of the Anopheles gambiae complex, Parassitologia, 10, 155

Coluzzi, M. et al. (1969) Cytogenetic observations on the saltwater species, Anopheles merus and Anopheles melas, of the gambiae complex, Parassitologia, 11, 177

Coluzzi, M. & Coluzzi, A. (1970) Incroci tra popolazioni di Antopheles labranchiae Falleroni, 1927 e Anopheles atroparvus van Thiel, 1927, Atti V. Congr. Soc. Ital. Parassitol.

Coluzzi, M. (1972) Inversion polymorphism and adult emergence in Anopheles stephensi, Science, 176, 59

Coluzzi, M. et al. (1972) Polimorfismo cromosomico e lunghezza dell'uovo in Antopheles stephensi, Parassitologia, 14, 261

Coluzzi, M. (1973) Laboratory and field observations on inversion polymorphism in anopheline mosquitos, Proc. 9th Int. Cong. trop. Med. Malar. (Athens), 250

Coz, J. (1973a) Contribution à l'étude du complexe A. gambiae. Répartition geographique et saisonaire en l'Afrique l'ouest, Cah. ORSTOM, ser. Ent. méd. et Parasitol., XI, 1, 3

Coz, J. (1973b) Contribution à la biologie du complexe A. gambiae Giles en Afrique Occidentale, Cah. ORSTOM, sér. Ent. méd. et Parasitol., XI, 1, 33

Crick, F. (1971) General model for the chromosomes of higher organisms, Nature, 234, 25

Davidson, G. & Jackson, E. (1962) Incipient speciation in Anopheles gambiae Giles, Bull. Wld Hlth Org., 27, 303

Davidson, G. (1964) The five mating types in the Anopheles gambiae complex, Bull. Wld Hlth Org., 43, 167

Davidson, G. & White, G. B. (1972) The crossing characeristics of a new, sixth species in the Anopheles gambiae complex, Trans. roy. Soc. trop. Med. Hyg., 66, 531

Davidson, G. & Hunt, R. H. (1973) The crossing and chromosome characteristics of a new sixth species in the Anopheles gambiae complex, Parassitologia, 15, 121

Davidson, G. & Zahar, A. R. (1974) The practical implication of resistance of malaria vectors to insecticides, Bull. Wld Hlth Org., (In press)

Dobzhansky, T. (1951) Genetics and the origin of species; 3rd ed'n., New York: Colombia University Press, 364 pp

Frizzi, G. (1947) Cromosomi salivari in Anopheles maculipennis, Sci. genet., 3, 67

Frizzi, G. (1949) Genetica di popolazioni in Anopheles maculipennis, Ric. Sci., 19, 544

Hunt, R. H. (1972) Cytological studies on a new member of the Anopheles gambiae complex, Trans. roy. Soc. trop. Med. Hyg., 66, 532

Ismail, I. A. H. & Hammoud, E. I. (1968) The use of coeliconic sensillae on the female antenna in differentiating the members of the Anopheles gambiae Giles complex, Bull. Wld Hlth Org., 38, 114

Jones, M. D. R. & Gubbins, S. (In press) Genetic differences in circadian flight activity in Anopheles stephensi

Kitzmiller, J. B. (1967) Mosquito cytogenetics. In: Wright, J. W. & Pal, R., 133

Kitzmiller, J. B. et al. (1967) Evolution and speciation within the maculipennis complex of the genus Anopheles. In: Wright, J. W. & Pal, R., 151

Kitzmiller, J. B. et al. (1973) Chromosomal differences between populations of Anopheles nuneztovari, Bull. Wld Hlth Org., 48, 435

Kreutzer, R. D. et al. (1972) Inversion polymorphism in the salivary gland chromosomes of Anopheles darlingi Root, Mosquito News, 32, 555

Macdonald, G. (1957) The epidemiology and control of malaria, London. Oxford University Press

Maynard Smith, J. M. (1966) The theory of evolution, 2nd ed'n. Hammondsworth: Penguin, 336 pp.

Paterson, H. E. et al. (1963) A new member of the Anopheles gambiae complex: a preliminary report, Med. Proc., 9, 414

Service, M. W. (1970a) Identification of the Anopheles gambiae complex in Nigeria by larval and adult chromosomes, Ann. trop. Med. Parasit., 64, 131

Service, M. W. (1970b) Ecological notes on species A and B of the Anopheles gambiae complex in the Kisumu area of Kenya, Bull. ent. Res., 60, 105

Service, M. W. (1972) Identification of the Anopheles gambiae complex in the Western Nyanza area, Kenya, 1971. WHO/VBC/72.362

Shidrawi, G. R. (1972) The distribution and seasonal prevalence of members of the Anopheles gambiae species complex (species A and B) in Garki district, Northern Nigeria. WHO/MAL/72.776

Srb, A. M. et al. (1963) General genetics. San Francisco: Freeman, 557 pp.

Swanson, C. P. et al. (1967) Cytogenetics. New Jersey: Prentice-Hall, 194 pp.

Wasserman, M. (1963) Cytology and phylogeny of Drosophila, Am. Nat., 896, 333

White, G. B. et al. (1972a) Comparative studies on sibling species of the Anopheles gambiae Giles complex (Dipt., Culicidae). Bionomics and vectorial activity of species A and species B at Segera, Tanzania, Bull. ent. Res., 62, 295

White, G. B. (1972b) The Anopheles gambiae complex and malaria transmission around Kisumu, Kenya, Trans. roy. Soc. trop. Med. Hyg., 66, 572

White, G. B. & Muniss, J. (1972c) Taxonomic value of spermatheca size for distinguishing four members of the Anopheles gambiae complex in East Africa, Bull. Wld Hlth Org., 46, 793

White, G. B. & Rosen, P. (1973) Comparative studies on sibling species of the Anopheles gambiae Giles complex (Dipt., Culicidae). II. Ecology of species A and species B in Savanna around Kaduna, Nigeria, during transition from wet to dry season, Bull. ent. Res., 63

White, M. J. D. (1961) The chromosomes. London: Mathuen, 188 pp.

White, M. J. D. (1973) Animal cytology and evolution. Cambridge

World Health Organization (1968) Cytogenetics of vectors of disease of man, Wld Hlth techn. Rep. Ser., No. 398

Wright, J. W. & Pal, R., eds (1967) Genetics of insect vectors of disease. Elsevier, Amsterdam, 794 pp

Zahar, A. R. et al. (1970) An attempt to group freshwater species of the Anopheles gambiae complex by some morphological larval and adult characters, Parassitologia, 12, 31

CHAPTER 2. ENTOMOLOGY IN MALARIA PROGRAMMES

CONTENTS

		Page
1.	INTRODUCTION	56
2.	TYPES OF ENTOMOLOGICAL ACTIVITY IN DIFFERENT EPIDEMIOLOGICAL SITUATIONS	59
3.	TECHNIQUES AND PROCEDURES FOR DIFFERENT TYPES OF ENTOMOLOGICAL ACTIVITY IN MALARIA PROGRAMMES	65
4.	ORGANIZATION OF ENTOMOLOGICAL ACTIVITIES IN MALARIA ERADICATION AND CONTROL PROGRAMMES	75
5.	ENTOMOLOGICAL STAFFING PATTERN IN MALARIA PROGRAMMES	78
6.	ENTOMOLOGICAL ACTIVITIES FOR THE PLANNING AND EVALUATION OF ANTI-LARVAL OPERATIONS	83
BIBLIOGRAPHY		94
ANNEX 1.	THE SELECTION OF INDICATOR VILLAGES AND CAPTURE STATIONS	95

CHAPTER 2

ENTOMOLOGY IN MALARIA PROGRAMMES

1. INTRODUCTION

The ninth report of the Expert Committee on Malaria (1962) clearly stated the main entomological activities to be carried out in different phases of malaria programmes. The guidelines for the types of entomological activity to be carried out in the various phases were so well described in this report that the appropriate paragraphs were reproduced almost in their entirety in the eleventh report of the Expert Committee on Malaria (1964) which dealt mainly with malaria entomology. In addition, the eleventh report described in detail situations where different methods should be used for the study of normal vector biology and the effect of impact of insecticides under different environmental conditions.

In brief, the entomological component in malaria programmes (whether of eradication or control) may be regarded as participating in the two main spheres of:

(a) planning for antivectorial measures and for their evaluation, and

(b) evaluation, consisting of general monitoring of the execution and impact of antivectorial measures, participation in epidemiological investigation of areas of persistent transmission and checking the receptivity and amenability to antivectorial measures in selected areas at intervals after eradication or after the required degree of control has been achieved. There should be a continuous feed-back from evaluation for planning purposes throughout the programme, and the evaluation methods themselves may be affected by this. As regards transmission, entomological data can only embellish or explain facts which are more easily and accurately determinable by parasitological survey. It is in the operational sphere that entomology can make its most important contributions, such as in the selection of the measure or combination of measures to be employed, insecticide of choice, duration of residual effect and spraying cycle, delimitation of spheres of vector influence, monitoring insecticide efficiency, etc.

1.1 PLANNING ACTIVITIES

These include:

Pre-operational	(a)	Review of existing epidemiological information and participation in preliminary surveys.
	(b)	Participation in the synthesis of the epidemiological data obtained in these.
	(c)	Formulation of the appropriate antivectorial measures to be employed.
	(d)	Preparation of the entomological staffing pattern on the basis of the envisaged evaluation procedures.
	(e)	Participation in drawing up the plan of operations.
Preparatory Phase	(f)	Establishment of field units and supervisory teams.
	(g)	Selection of indicator villages.
Later Phases	(h)	Joint entomological/parasitological assessment of attack measures.

1.2 EVALUATION ACTIVITIES

Preparatory Phase   (a)   Acquisition of baseline data for assessment of impact of attack measures and for checking efficiency of the particular antivectorial measures to be employed.

Attack Phase   (b)   Monitoring efficiency of antivectorial measures.

(c)   Investigation of refractory situations.

(d)   Advice on modification of attack measures or special remedial measures.

(e)   Review of any side effects of insecticide application on silkworms, bedbugs, sandflies, etc.

Consolidation Phase   (f)   Monitoring of receptivity status of vulnerable areas.

(g)   Checking amenability to re-imposition of antivectorial measures.

(h)   Investigation of persistent foci.

(i)   Checking efficacy of remedial measures adopted.

Maintenance Phase   (j)   Periodic checks of receptivity and amenability to antivectorial measures in selected areas.

1.3 BASIC PRINCIPLES FOR PLANNING AND CARRYING OUT ENTOMOLOGICAL ACTIVITIES

1.3.1 Epidemiological

The entomological activities will be planned and carried out in representative areas with different patterns of distribution of the disease. The dynamics and behaviour of the vector population, and the role played in the transmission of malaria will be studied in relation to breeding potentialities, contact with man, and changes in the human malaria reservoir.

The epidemiological information should be available at all stages as it is required for initial planning and subsequent changes in the programmes as it progresses. The essential requirements are as follows:

(a) In the initial stages it would be necessary to map out epidemiologically important areas (from preliminary epidemiological/entomological surveys if sufficient information is not available) in order to draw up a plan for applying attack measures and a method of epidemiological evaluation. This mainly determines the scale of the entomological activities and consequently the staffing pattern.

(b) As a programme progresses, there should be close liaison between the epidemiological, entomological and operational staff to ensure rapid exchange of information and consequently timely direction of entomological activities.

(c) There should be a continuous joint analysis and interpretation of the entomological and parasitological data as integral parts of the epidemiological evaluation, on the basis of which modification of entomological activities can be introduced as the programme develops.

1.3.2 Ecological

The following points are recapitulated from Chapter 1, and should be kept in mind when planning:

(a) Each vector species has a more or less specific life history and behaviour related to specific habitats for larvae and adults determined by the geographical and ecological conditions.

(b) The life cycle and behaviour of the vector population varies between certain limits under the influence of the biotic and abiotic factors such as:

- macro and micro-climatological factors (temperature, humidity, rain, wind, light);

- presence and proximity of favourable breeding and resting places;

- presence and number of different hosts, and habits of hosts;

- the modification of the environment produced by man, e.g., use of insecticides, environmental manipulation, source reduction, land development schemes, etc.

In the light of these aspects it is easy to understand that the same vector species might show different behaviour during different seasons or in different environmental situations.

1.3.3 Implementation

When entomological evaluation is to be carried out and once the objectives have been clearly defined, the following details must be considered:

(a) Methods and their application. The techniques should be selected according to objectives, local conditions and behaviour of the vector. Selection of the spots to be investigated. Intensity of sampling. Time and duration of the evaluation.

(b) Sampling. As has been mentioned before, this should be representative of the investigated areas.

(c) Processing. The processing of collected mosquitos is extremely important and different procedures are employed according to the objectives of the sampling.

1.3.4 Logistics

On the basis of the aims, scale of application of techniques to be employed and the frequency of entomological observations, estimates of staff required and other facilities can be made to ensure the following:

- adequacy of trained personnel of different categories,

- adequate supplies and equipment (check lists for each entomological observation should be established),

- adequate transport to ensure timely movement of the teams,

- incentives for night work,

- facilities for recording and rapid consolidation and presentation of data.

1.3.5 Work schedules

Schedules for each entomological team should be developed. Depending on the types of activity at each stage, a weekly itinerary should be worked out, defining the exact timing and sequence of observations in areas assigned to each team.

1.3.6 Training and supervision

Basic and refresher training should lead to the development of high proficiency in performing the entomological techniques.

Post descriptions for each category of personnel should be established, also the responsibilities of entomological teams at various levels should be clearly defined.

Supervision should be effected by:

- surprise visits,

- critical appraisal of the results contained in reports of the entomological teams,

- supervisors actually participating in the work of the teams, detecting and correcting mistakes on the spot.

2. TYPES OF ENTOMOLOGICAL ACTIVITY IN DIFFERENT EPIDEMIOLOGICAL SITUATIONS

At present, the situation of antimalaria activities can be summarized as follows. There are:

(a) areas under maintenance phase, or areas where malaria has been eradicated. Foci of malaria transmission in some programmes has necessitated reversion to the attack phase in parts of areas under maintenance. In some countries where malaria eradication has been achieved, some recurrence of malaria transmission has occurred in limited areas;

(b) areas under consolidation (in some of which it has been necessary to revert partly or totally to attack);

(c) areas with prolonged attack using DDT or an alternative new insecticide;

(d) areas under attack measures for malaria control or for maintaining the previous gains, with malaria eradication being the long-term objective;

(e) areas where malaria control has been started;

(f) areas where there are no antimalaria activities but where malaria control is envisaged.

With this diversity of situations which will also be changing over the years, the aspect of organization of entomological work in malaria programmes has to be emphasized in this manual in such a way as to ensure a flexible approach to cope with these different situations in programmes which would be applicable for many years to come.

The best approach seems to be first to define the types of entomological activity (rather than techniques to be adopted) and outline why (i.e., the objectives), where and when they should be applied.

Long experience in antimalaria programmes whether originally for time-limited eradication or for control makes it possible to categorize entomological activities into the following types, the definition, objectives and applicability of which are shown in Table 2:

(a) PRELIMINARY SURVEYS. These are short-term surveys aimed at the production of information for planning purposes.

(b) REGULAR OR TREND OBSERVATIONS. These are also termed "routine" or longitudinal observations.

(c) <u>SPOT-CHECKS</u>. These are sometimes called "random" surveys or surveys at selected localities.

(d) <u>FOCI INVESTIGATIONS</u>. Investigations in new or persisting foci of malaria transmission.

(e) <u>VIGILANCE</u>. This type of activity has been prescribed for areas under maintenance or where malaria has been eradicated.

Once the types of activity have been designated to deal with certain epidemiological situations then comes the time for the selection of the appropriate entomological techniques to be employed and procedures to be followed. These have been outlined below in section 3 and it is according to the local situation that the choice of these techniques and procedures has again to be adapted.

The entomologist and the malariologist, being aware of the epidemiological development in different areas in their programme, will therefore be able to select the type or combination of types of activity and the techniques that would be appropriate for the particular situation in each area.

On this basis the plan of work can be prepared, ensuring adequate staffing and facilities since the actual needs have been monitored.

For example:

(a) in areas under prolonged attack it is necessary to implement:

- trend observations in a few selected indicator villages,

- spot-checks,

- foci investigation, as they may arise;

(b) in areas under consolidation with persisting foci of transmission it is necessary to implement:

- short-time foci investigations for determining the reasons for persistence of transmission,

- trend observations in selected foci for a certain period to assess the impact of remedial action until transmission is eliminated,

- spot-checks in the remaining area under consolidation showing favourable results;

(c) in areas under maintenance with parts reverted to attack it is necessary to implement:

- trend observations in selected indicator villages for assessing the impact of the original insecticide used (e.g., DDT) or an alternative insecticide as a trial,

- spot-checks,

- foci investigation, as they may arise;

(d) in areas under maintenance, including those parts which have not been reverted to attack, it is necessary to implement:

- vigilance by spot-checks,

- a seasonal trend observation for determining vector potentials in selected localities in receptive and vulnerable areas.

Another example is a new malaria control programme employing residual house spraying and other measures. In such a programme it is necessary to implement:

- a preliminary survey,

- trend observations in selected indicator villages to establish baseline data,

- trend observations to assess the impact of the designed dosage and frequency of application of the insecticide or larvicide,

- trend observations in a few carefully selected indicator villages to ascertain that impact of attack measures continues to be successful in achieving, interrupting, or reducing malaria transmission to the level at which malaria is no longer a public health problem. Trend observations can be terminated as soon as results show favourable response to attack measures,

- spot-checks to be increased as trend observations are reduced or terminated. Spot-checks are to continue as long as attack measures are applied for maintaining the gains,

- foci investigations as the need arises in a prolonged attack. The investigation may show the need to modify the attack measures.

From the above examples it can be seen that according to need the types of activity can be selected and subsequently the manpower and transport and other facilities required, the formation of teams, their deployment for undertaking a single or a combination of entomological activities can be planned on a factual basis.

In fact, these steps are necessary in order to establish an efficient entomological service ensuring a flexible mobilization of the entomological teams for certain types of activities appropriate to the development of the epidemiological situation.

Entomological conclusions arrived at in one area should not be extended to other areas in which the environmental factors are different. The area of distribution of a species may be a very large one and the phenology and behaviour of different populations of the same vector may present differences due to the influence of the various factors mentioned in section 2 of Chapter 1. Thus the entomological activity is a complicated one and requires not only the scientific application of entomological methods but also a careful recording of environmental factors which may vary from place to place, from season to season and even from day to day. In this situation it is easy to understand that generalization and theoretical calculations based on a few small and restricted sampling occasions should be avoided. The entomological component of a malaria programme has a flexible character and should be modified according to the objectives of the activities of the particular phase of the programme and measures employed.

3. TECHNIQUES AND PROCEDURES FOR DIFFERENT TYPES OF ENTOMOLOGICAL ACTIVITY IN MALARIA PROGRAMMES

3.1 PRELIMINARY SURVEY

A preliminary survey should provide the following information for each locality investigated:

- The anopheline species detected.

- Rough relative adult densities —

    (a) resting (hand collection or pyrethrum),

    (b) biting (man and animal),

    (c) exit trap (where houses are suitable and time permits).

- Distance from larval sites, and type of latter (sample of larvae should be taken for confirmation of species present).

- Seasonal accessibility of locality.

- Rough indication of malaria prevalence (complementary parasitological survey should be carried out in the same locality).

- Notes on human ecology, population size, population movement, domestic animals, etc.

- The history of insecticide treatment, whether applied for public health or agriculture.

3.1.1 Techniques and procedures

Preliminary surveys are carried out as follows:

(a) Existing information on the area is assembled and analysed.

(b) A rough delimitation is made of the different geographical areas and environmental conditions, and the area is divided into more or less homogeneous zones from the point of view of the vector (if information is available) and malaria prevalence.

(c) The mosquito production potential is estimated by making a reconnaissance of the different types and surface areas of water (actual and/or potential) within each zone.

(d) The most appropriate method for sampling mosquito species is selected. If the habits of the local species are unknown insofar as resting behaviour is concerned, larval collection should be applied in combination with adult sampling. When resting behaviour is known (as is obviously now the case for all recognized malaria vectors) adult collection should be the method of choice for indoor resting mosquitos. Night bait and trap collections are employed for exophilic mosquitos.

(e) A number of sampling localities for collection of adult mosquitos are selected in each geographically homogeneous zone, taking into account distance from the larval sites (when known). At least 9-10 houses are sampled in each locality with endophilic vectors, and at least two spots (indoor and outdoor) are selected for night bait collection among those houses where the highest resting density has been observed.

(f) Preliminary susceptibility tests should be carried out.

The number of localities to be sampled and the amount of work to be carried out cannot be given in a ready-made formula and the entomologist has to exercise his judgement in order to obtain a satisfactory and comparable sample. Two or three localities for each representative area might prove satisfactory, but in general four would be recommended.

The time of sampling is very important. The survey should be carried out subsequent to the malariometric survey (at least a spleen survey) but during the most favourable period of mosquito production, preferably during the expected peak density period. The survey should be completed in the shortest possible time to avoid changes in the environmental conditions which have a bearing on the abundance of the mosquito population. It should not be spread over more than two months, depending on the area. If the investigator does not know the period of expected peak density this can be estimated by ascertaining the seasonal prevalence of malaria. In general, the peak density of mosquitos precedes the peak of malaria incidence by about one month.

In areas where two vectors are present the survey should be carried out during the peak density of the most important vector, followed by a survey during the peak densities of other vectors. When performing night bait catches during a preliminary survey, sampling in the first or latter half of the night is usually satisfactory for most species of mosquito, though this should be confirmed initially by one or two all-night catches. Such a collection will include all early- and late-biting species under usual circumstances. Light-trap catches could be used as an additional method of collection.

The information required from a preliminary survey, the methods employed and suggested sampling procedures are given in Table 3.

3.1.2 Synthesis of data

On the basis of a survey of previously existing information and results obtained by preliminary entomological and parasitological surveys, preparations are made for more detailed sampling for the acquisition of baseline data as follows:

(i) Define the objectives for the collection of entomological data.

(ii) Identify the different ecological conditions and the topographical distribution of vector species and incidence of the disease.

(iii) Select representative localities for each ecological area on the basis of the following information:

- incidence of the disease,

- types and distribution of larval sites and production potential,

- types of human habitats, customs and occupations,

- domestic animals,

- all year round access.

(iv) Select capture stations in each locality, bearing in mind the following aspects:

- type of human habitation,

- distance from larval sites,

- presence or absence of domestic animals in or near the compound to be investigated,

- behaviour of the vector.

TABLE 3. PRELIMINARY SURVEY

Information needed for each locality	Methods employed[a]	Suggested sampling procedure		
		Place	Time	Frequency
Rough indication of malaria prevalence	1. Consult local records for the results of the malariometric survey. Examine sample of population for malaria parasites if information is insufficient	Two localities for each type of ecological area	In course of survey	Once
Anopheline species present and their distribution (geographical and ecological)	2. Larval collection	Two localities for each type of ecological area. In each locality 5 - 10 breeding places for each type of breeding site	Daytime, during assumed breeding season	Once
	3. Pyrethrum spray or hand collection	At least in the same two localities for each type of ecological area as mentioned above. In each locality 10-15 indoor resting places representing different types	Between sunrise and two hours before sunset during assumed season of high density	Once
	4. Night-biting catch on human or animal bait	Two spots in each locality	In the first or latter half of the night depending on the results of trial all-night catches	Once
	5. Light-trap collection in association with use of bed-nets	Two spots in each locality	All night	Once
Rough adult relative densities	As 3 above	4-5 localities with high malaria prevalence. In each locality 10-15 resting places as close as possible to breeding sites	Daytime. During season of expected high anopheline prevalence	Two to three times at 2-4 week intervals within two months
	As 4 above. Human bait	Two spots in each of above localities	The first or latter half of the night. During season of expected high anopheline prevalence	Two to three times at 2-4 week intervals within two months
	6. Exit-trap collection if time permits	Three to five houses in each locality	All night. During season of expected high anopheline prevalence	Two to three times at 2-4 week intervals within two months
Distance from larval sites, and type of latter	7. Direct observation	Each locality surveyed or selected on sample basis for this observation	In course of survey	Once
Seasonal accessibility of locality	8. Local enquiry, state of roads, bridges, etc.	Each locality surveyed or selected on sample basis for this observation	In course of survey	Once
General information on human ecology	9. Notes on population size, movement, domestic animals etc.	Each locality surveyed or selected on sample basis for these observations	In course of survey	Once

[a] The material collected should be processed as follows:
    (a) identification of vectors and other anopheline species;
    (b) classification of the blood digestion stage of samples of vectors found in day-time resting places;
    (c) dissection for gland dissection of samples obtained in day-time and bait capture if vectors are not known and time permits.

On the result of the above preliminary entomological survey plus the result of malariometric survey and budgetary considerations, it will be decided:

(a) whether the antimalaria measures will be based on anti-vector operations alone or in combination with other preventive measures, e.g. mass drug administration, anti-relapse treatment, that is, differential measures according to local epidemiology;

(b) what type of anti-vector measures will be employed, antilarval or imagocidal or both;

(c) what method and frequency of application.

The baseline data to be longitudinally collected, as will be shown later, will make it possible to adjust the plan of attack measures.

3.2 TREND OR REGULAR OBSERVATIONS FOR BASELINE DATA AND EVALUATION OF ATTACK MEASURES

The techniques and procedures given in this section are principally meant for entomological evaluation of attack measures by residual house spraying. Those pertaining to the evaluation of anti-larval measures are given in Section 6 of this Chapter.

3.2.1 Localities of observation

The trend observations should be conducted in selected villages, termed indicator villages, in which capture stations having the highest density should be chosen. The indicator villages should be part of the localities where parasitological surveys are undertaken.

The selection of the indicator villages should take into account the presence of high malaria endemicity, the topography of the area, the presence of high vector density in the area of its dominance, the seasonal distribution of the vector and the accessibility of these localities throughout the year.

Full details are given in Annex I to this Chapter.

3.2.2 Techniques and procedures

The techniques and procedures to be used are illustrated in Table 4. They include:

(a) Hand spray capture, night observations and man-bait capture indoors and outdoors at fortnightly intervals, if facilities permit, otherwise at monthly intervals.

(b) Window-trap observations should be conducted for at least 2-3 nights every two weeks. The number of trap premises during the pre-spraying period can be limited to 3-5 in view of the high vector density prevailing. These should be increased after application of spraying to 10 traps.

(c) If the abdominal stages of house-resting and window-trap collections indicate exophilic tendencies, outdoor searches in natural shelters should be attempted as a special investigation until the concentration sites are located. The outdoor collection should be done periodically and not routinely with the aim of attempting to establish the level of density before and after spraying and for collecting a large sample of bloodmeal smears for precipitin testing.

(d) Susceptibility tests should be carried out to establish baseline data at different seasons during the preparatory phase. After spraying, periodical checks should be made before the application of the first spraying round and prior to each subsequent round.

TABLE 4. TREND OBSERVATIONS FOR BASELINE AND FOR EVALUATION OF RESIDUAL HOUSE SPRAYING

Information needed	Methods employed	Suggested sampling procedure Place	Suggested sampling procedure Time	Suggested sampling procedure Frequency
Anopheline species present and their distribution (geographical and ecological)	1. Information from preliminary survey. Continually brought up to date			
Vector relative densities and behaviour	2. Indoor pyrethrum spray, or hand collection in special situations	Ten fixed capture stations in each chosen locality	Daytime, preferably early morning	Once every two weeks or every month according to facilities
	3. Exit-trap catch, and after spraying 24 hour survival	3-5 rooms per locality before spraying, 10 rooms per locality after spraying	All night	2-3 nights every two weeks or every month according to facilities
	4. Resting and making exodus in traps: classification of abdominal stages of material obtained under 2, 3, and 6	As 2, 3, and 6	As 2, 3, and 6	As 2, 3, and 6
	5. Biting: man-bait collection indoor and outdoor	Collection in two stations with a minimum of two baits indoor and two baits outdoor but attempts to use volunteers to increase the number of baits should be made	All night	Once every two weeks or once every month according to facilities
	6. If indicated from 4, outdoor, hand collection and/or artificial shelter collection of outdoor resting mosquitos	Highest possible number of natural shelters and about 10 artificial shelters in each locality selected for outdoor sampling	Daytime, preferably early morning	Periodically as a special investigation
	7. Host preference: precipitin testing of bloodmeal smears	Material from 2 and 6 above in two or three localities in each ecological area as a special investigation	As 2 and 6 above	Once or twice a year in season of vector prevalence
Vectorial importance	8. Salivary gland dissection of material obtained by 2, 3, 5, and 6, but not after spraying except in special circumstances (see text)	As 2, 3, 5, and 6 above	As 2, 3, 5, and 6 above	As 2, 3, 5 and 6 above
	9. Determination of parous ratio for longevity estimate by dissection of material obtained by 2, 3, 5, and 6 if possible, otherwise 5 only (see text)	As 2, 3, 5, and 6 above	As 2, 3, 5, and 6 above	As 2, 3, 5, and 6 above
Susceptibility levels of selected anopheline species to candidate insecticides or the insecticide that is already in use	10. WHO standard susceptibility test	Two localities in each ecological area, including the indicator villages	Season of maximum anopheline availability	Once to establish baseline, thereafter every six months using discriminating concentration. After spraying, before the application of each spraying round

3.2.3   Notes on processing of material

(a)   Classification of the blood digestion stages of samples of vectors and suspected vectors found in day-time resting places indoors and outdoors and those taken leaving by window traps;

(b)   Dissection for gland infection and the parous rate determination of samples of vectors and suspected vectors collected from man-bait capture. Staff is especially trained to undertake the simplified Polovodova's technique. Dissection may be extended to samples collected from indoor resting shelters and to those collected from outside resting shelters should the results of precipitin tests indicate that a proportion of vectors had fed on man. After spraying no sporozoite-positive mosquitos are usually found, thus gland dissections should be excluded except in the case of an efficient vector such as A. gambiae.

3.3   SPOT-CHECKS

Trend observations carried out in fixed indicator villages may indicate a good response, and entomological investigation in foci of transmission will indicate the reasons for a breakdown which has already occurred. Therefore, spot-checks are required to cover other areas in order to detect places where there is vector potential, thus providing a priori evidence of abnormal happenings in vector populations in areas to be suspected on epidemiological grounds whether under attack operations, consolidation or maintenance or where malaria has been eradicated.

Since the entomological activities cannot be organized on total coverage basis, localities to be surveyed by spot-checks should be selected jointly by the malariologist, entomologist and the operational officer according to the following criteria:

(a)   areas with high breeding potential, such as those affected by heavy rainfall or those with permanent water impoundment or with man-made breeding places;

(b)   areas suspected for possible deficient spraying operations in areas under attack;

(c)   areas to be suspected on the basis of malariogenic potential and areas vulnerable to influx of sources of infections.

3.3.1   Techniques and procedures

In areas where residual house spraying is being applied, the entomological team should first delimit the new, missed and disturbed premises, which should as far as possible all be surveyed. In addition, a sample of the sprayed premises should also be searched. Usually a single technique is employed in this type of survey. Search by hand/spray capture is normally utilized with endophilic species, but bait capture should be utilized with exophilic species such as A. balabacensis and A. nuneztovari. Indoor and outdoor sites for bait capture are to be selected where houses have been discovered with operational deficiencies as well as houses which are or appear to have been properly sprayed.

In areas under consolidation, maintenance or where malaria has been eradicated, the target would be the houses near breeding places whether natural or man-made.

3.3.2   Processing of material

The samples of vectors collected should be dissected for parity and for infectivity where a high density is encountered.

Results of spot-checks should be reported immediately to the epidemiological and operational sections for deciding what action should be taken.

3.4 ENTOMOLOGICAL INVESTIGATIONS IN FOCI OF TRANSMISSION

Information indicating the need for entomological investigation must be rapidly transmitted from different zones to the entomological teams in charge so that a timely investigation can be carried out before the intervention of seasonal changes occurs or focal spraying is applied as a remedial action.

3.4.1 Localities of investigation

In planning this activity it should be borne in mind that the entomological manpower cannot deal with all cases of malaria appearing in the area. The time and efforts of the entomological teams should not be wasted on scattered indigenous cases while there is evidence of more significant transmission occurring in certain foci. For example, the entomological investigation should be primarily directed to localities which show clusters of indigenous cases or infant positive cases or localities with P. falciparum infection. If the number of such foci is high, the investigations should be concentrated only in a few representative foci in each area.

Once the reasons for persistence of transmission have been demonstrated in a focus representing a homogeneous ecological substratum, there is no point in duplicating investigations in other areas with similar conditions.

The team should be directed to work in foci in other ecological areas as well as to carry out follow-up observations in foci which have been subjected to remedial action.

3.4.2 Procedures

The entomological investigation in a focus of malaria transmission requires close collaboration between the epidemiological, entomological and operational staff.

Reinvestigation of the cases may need to be carried out by the epidemiologist and further fever survey may need to be applied on collaterals.

The operational officer should assist in delimiting the operational defects in the entire locality while the entomologist and his staff concentrate on the houses of the indigenous cases and surrounding houses. Flexibility should be exercised in selecting the techniques. There are no standard procedures for such an investigation. However, the following procedures have been found appropriate (see Table 5):

(a) the history of each indigenous case should be checked and the approximate date of the primary attack should be recorded in order to see the possible date of contraction of infection in relation to the date of spraying as illustrated in Fig. 16;

(b) the cases should be located on a sketch map. If their number is high, select groups concentrated in clusters;

(c) upon arrival at the locality, the entomological team should observe the quality of spraying and conditions of coverage in the patients' houses and the surrounding houses;

(d) as vector density may be low, it is advisable to avoid large-scale spray capture. Only visual observations should be made of vectors resting in patients' and adjacent houses. Where some density is found further enquiries should be made to confirm from inhabitants whether the premises were sprayed as it is often difficult to observe the deposits on certain surfaces;

(e) outlet window traps should be installed in sprayed premises to determine the mortality of vectors. This will be particularly important in cases of vector resistance to the insecticide. Trap observations can also be conducted in disturbed and unsprayed premises for comparison;

TABLE 5. INVESTIGATION IN A FOCUS OF PERSISTING OR RENEWED TRANSMISSION

| Information needed | Methods employed | Suggested sampling procedure |||
		Place	Time	Frequency
Is insecticide present and properly applied?	1(a) Examination of spraying records and visual inspection of deposits in sprayed structures	Positive locality		
	1(b) In some localities, where test insects are available, bioassay may be carried out if time permits	Positive house and surrounding houses. Five cones each	Early morning	Once
What vectors and other anophelines are present — their relative density and contact with man?	2(a) Night-biting collection, indoor and outdoor human bait	Vicinity of positive cases	All night	Once. Repeat if required
	2(b) Night catches round animal bait in open or in cattlesheds if unsprayed	Vicinity of positive cases	All night	Once. Repeat if required
	3. Exit-trap catch, with 24-h survival rate	At least five houses, including those with positive cases	All night	Duration of investigation
	4. Pyrethrum spray or hand catch indoors, hand catch in possible outdoor shelters	Positive locality	Daytime, early morning	Once
	5. Larval survey if indicated (see text)	Radius of 500 m round positive locality	Daytime	Once
Is transmission still persisting?	6(a) Salivary gland dissection of all parous anophelines collected in the focus, or a good sample of these	Positive locality, especially from houses of positive cases	As convenient	Throughout investigation
	6(b) Parous rate determination on samples collected biting man			
Are vectors still susceptible to insecticides used?	7. Susceptibility test using discriminating concentration, if mosquito material available	Positive locality or vicinity. Bred-out material may be used	During investigation	As required by availability of material

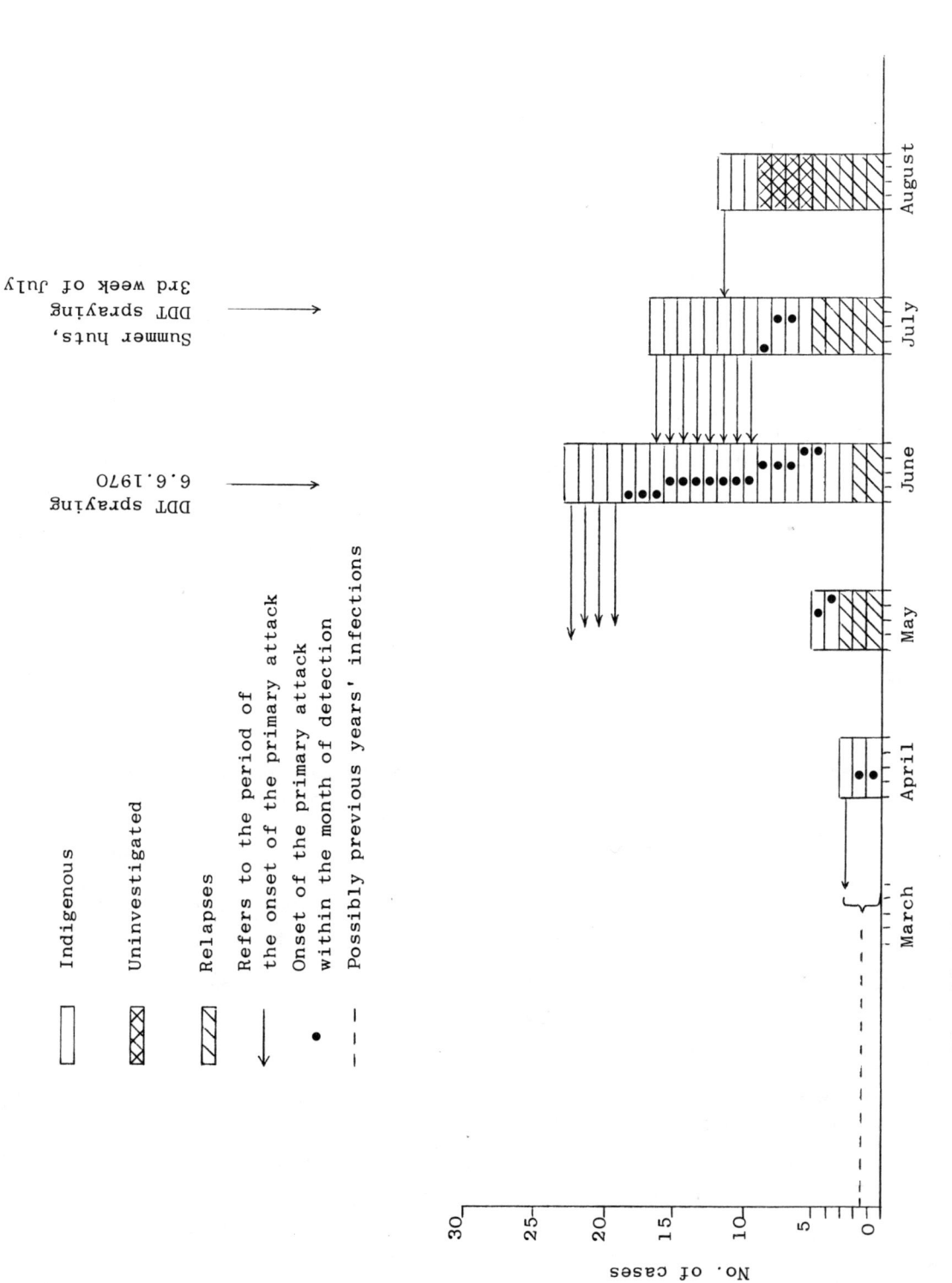

Fig. 16 Parasitological situation, Khumdan, Afghanistan, 1970.

Full acknowledgement is made to the Malaria Institute, Kabul, Afghanistan for providing the parasitological data collected by the Northern Region, Kunduz Office, 1970.

(f) a survey should be made to show the distribution of the inhabitants between indoors and outdoors during the night in order to determine the risk of exposure to vector biting in the two sites;

(g) a survey of the ecological conditions should be conducted in conjunction with sampling of bloodmeal smears, particularly in the case of vectors exhibiting exophilic tendencies;

(h) bait capture on man indoors and outdoors should be conducted for the whole night. Samples should be dissected for parity and infectivity determination.

(i) hand/spray capture should be applied in the morning of the last day of the investigation in as many premises as possible including those found to be newly constructed or which remained unsprayed or were altered after spraying.

(j) additional observations if:

- a vector is found only by bait capture - searches in outside shelters should be made and bloodmeal smears collected from the outside resting population;

- there is an indication of incipient or vector resistance to the insecticide or a certain level of resistance has been recorded elsewhere - susceptibility tests should be conducted using the discriminating dosage. Information on the insecticides used in agriculture should be collected;

- larval searches would be indicated if no, or only a scanty, adult density of the vector was observed by other techniques.

3.5 VIGILANCE

3.5.1 Spot-checks

Localities to be surveyed

The localities to be surveyed by spot-checks for detecting the geographical distribution and density of various vectors should be selected according to the following criteria:

(a) representation of different geographical areas and varied ecological substrata;

(b) priority to be given to areas with high malariogenic potential, taking into account the following:

- the degree of previous stability of malaria

- its endemic level and seasonal changes

- areas with history of epidemics

- refractoriness or responsiveness of malaria to control measures and the previous susceptibility and behaviour of vectors

- area affected by great environmental changes (whether natural or man-made) including urbanization, land development, etc.

- the areas of high vulnerability and the time of the year influx of sources of infection occurs.

Techniques and procedures to be adopted

As described in section 4 of this Chapter.

3.5.2　Seasonal trend observations

Localities and timing of observations

From the results of spot checks, indicator localities should be selected in highly receptive and vulnerable areas, preferably where previous data are available from the period of pre-eradication and during the application of attack measures.

The trend observations should cover the season of vector high prevalence.

Techniques and procedures

(a)　House-resting density for previously endophilic vectors to determine whether it has regained its previous level.

(b)　Man and animal bait capture to determine whether the degree of contact with man remained as before or whether deviation to animals has occurred under new ecological changes.

(c)　Collections of bloodmeal smears from vectors of different biotopes for precipitin tests as additional support to observations under (b).

(d)　Susceptibility testing for determining any changes in vector reaction to the insecticide(s) which had been used and the newer insecticide that may need to be used for applying preventive or remedial measures.

In cases where a vector has apparently disappeared, this has to be confirmed by long-term extensive observations in different seasons, particularly in localities where the vector had been abundant, utilizing different techniques including larval surveys as in the following:

(a)　Total absence of the vector resting in houses by pyrethrum spray catches to be made on a wide scale.

(b)　Total absence of the vector in outside-resting shelters using several sampling devices.

(c)　Total absence of vectors biting man and animals using direct capture, baited nets and light traps in conjunction with baits protected with bed-nets. The catches must be extended to extradomestic sites, isolated hamlets that could have escaped previous control measures and the available forest area.

(d)　Total absence of immature stages of the vector by extensive as well as intensive larval searches. Careful identification of all larval specimens collected must be made since the presence of great abundance of non-vector anopheline species may mask a small number of larvae of the vector.

Raising adults from larvae will be a double check since some species are closely related in larval morphological features, also in view of the presence of early instars of different species which can hardly be differentiated.

4. ORGANIZATION OF ENTOMOLOGICAL ACTIVITIES IN MALARIA ERADICATION AND CONTROL PROGRAMMES

As mentioned earlier, the aims and types of the entomological work to be carried out in a time-limited malaria eradication programme are well-known as they have been clearly emphasized phase by phase in the ninth and eleventh reports of the Expert Committee on Malaria (1962) and (1964), respectively, and other documents. It is not intended here to repeat what has been given in these documents. In section 3 (above), the types of entomological activities as appropriate to the different epidemiological situations presently prevailing in malaria

programmes have been described. It is sufficient here to give an idea of the order of
magnitude of each type of activity throughout the phases of a typical time-limited malaria
eradication programme. This can be clearly explained in terms of the distribution of time
of the entomological staff on various activities in the different phases of an eradication
programme (see Fig. 17).

Consideration is now given to the organization and staffing pattern for entomological
work in a malaria control programme.

In a time-limited malaria eradication programme, the overall function of the entomological
activities is to provide, together with the corresponding parasitological observations,
elements for an integrated epidemiological evaluation.

In a malaria control programme employing anti-vector measures, this function remains the
same since the epidemiological evaluation of the progress is likewise required and entomological findings form an integral part of it. The known difference lies in that a time-limited
eradication programme aims at complete interruption of malaria transmission and elimination of
the reservoir of infective cases, whereas the control programme applying anti-vector measures
is intended to reduce malaria endemicity to the extent that malaria no longer becomes a
significant public health problem. The entomological findings can, therefore, contribute to
the epidemiological assessment, showing whether declining trends in malaria transmission are
resulting from the application of attack measures. Furthermore, entomological investigations
may also be required for elucidating the factors hampering the progress in parts or the whole
area under the control programme.

The obvious difference may be the scale of the entomological operations which will depend
on the size of the areas designated for control operations which are usually much smaller than
in a country-wide malaria eradication programme. These areas are primarily selected according
to priorities for epidemiological reasons and land development.

From the above, it appears that in a malaria control programme employing anti-vector
measures, it is important to maintain certain types of activity but it is advisable to reduce
the techniques to be employed for assessing the impact on the vector population. The
entomological techniques _per se_, as will be discussed later, are not sensitive, besides which
the limitations in scale of application in some malaria eradication programmes have led to a
poor entomological contribution to the epidemiological evaluation. In fact, in malaria
control there will be a need to employ various types of activity in view of the possible use
of new insecticides or combined attack measures involving the use of larvicides and biological
control. On the other hand, some readjustment of the activities would need to be made when
the epidemiological target of the control measures has been achieved. For example, it may be
possible in such cases to limit the entomological activities to vigilance surveys in some
areas where near complete interruption of transmission has eventually been achieved. In the
following section, the essential entomological activities are discussed.

4.1   ENTOMOLOGICAL ACTIVITIES IN MALARIA CONTROL PROGRAMMES

The extent of the entomological activities in a malaria control programme will depend on
the types of control measure to be adopted as outlined in the following:

- Biological anti-larval measures.

- Larviciding.

- Source reduction measures.

- Residual insecticides.

4.1.1   Biological anti-larval measures

Spot checks on the presence of the biological agent and mosquito larvae and adults. The
simple method of larval sampling, e.g., dipping, and adult sampling by partial or standard
spray-capture should be used with endophilic species but with the use of human and animal baits
in case of exophilic species.

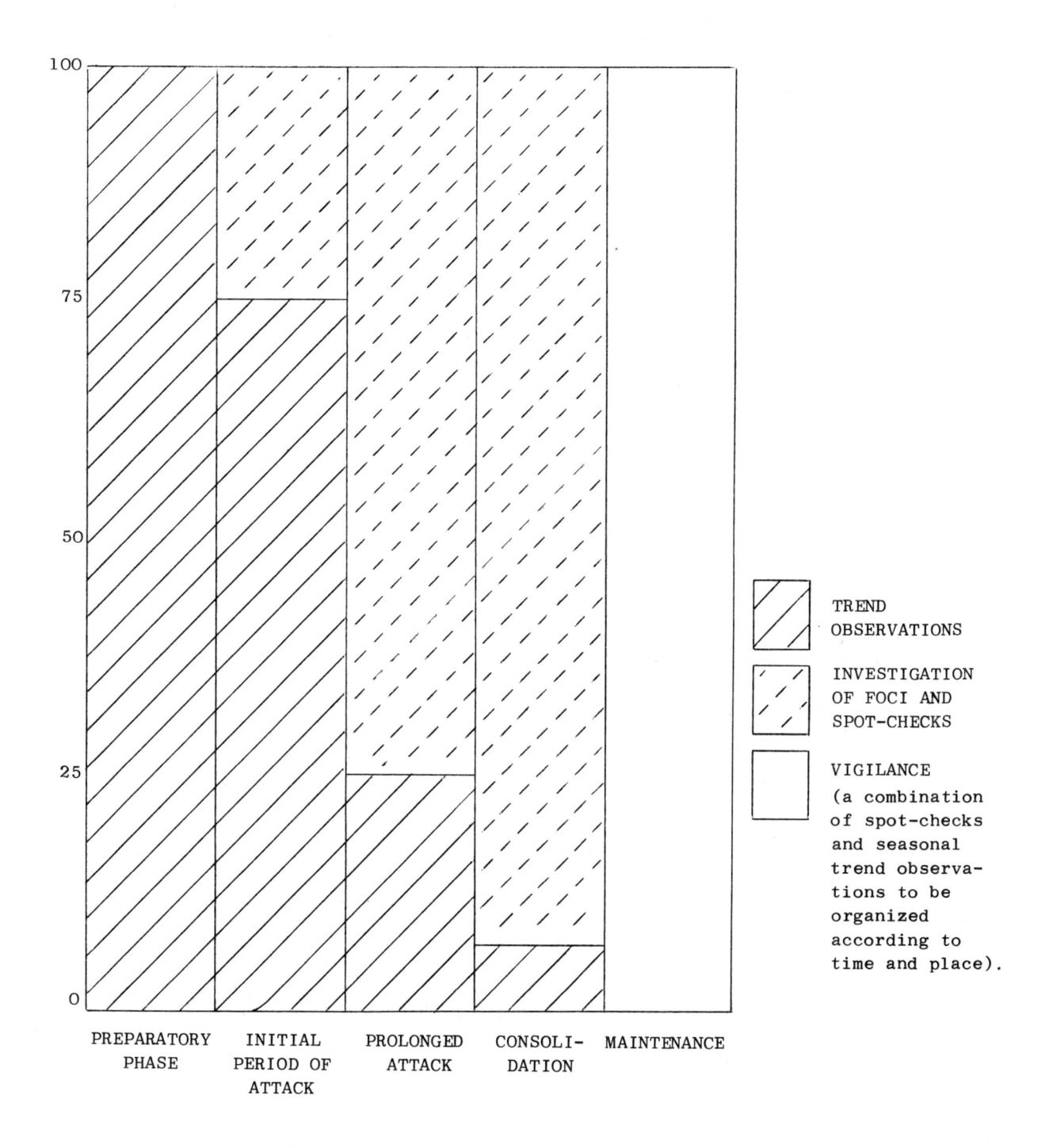

Fig. 17. A schematic diagram showing the relative distribution
of the working time of entomological teams among the different
activities according to the development of the programme

4.1.2   Larviciding

Regular larval sampling will be required and will need to be in phase with application frequency. The sampling should preferably be made near to the day of application of the insecticide. This type of regular larval sampling may be needed at the beginning of the campaign until it is confirmed that dosage and frequency of application prove effective in all types of breeding places.

Spot checks need to be made both for the detection of untreated and newly-established breeding places and for the presence of adult mosquitos in the way described above. Additionally, periodical larval susceptibility testing will be required.

4.1.3   Source reduction

Entomological activities should be largely of the vigilance type involving detection of breeding places and searching for adults.

4.1.4   Residual insecticide

In order of priority, the following entomological techniques need to be employed:

- Quantitative estimation of man-vector contact.
- The direct assessment of mosquito mortality due to insecticides by window-trap observations.
- Parous rate determination.
- Sporozoite rate determination, if feasible.
- Spot checks by partial or standard spray-capture where endophilic vectors are involved or by limited-time human-bait catches where the vector is largely exophilic.

Periodical adult susceptibility testing, though not included in this priority list, is obligatory.

Adequate baseline data should be collected appropriate to particular control measures to be used.

5. ENTOMOLOGICAL STAFFING PATTERN IN MALARIA PROGRAMMES

Entomological data can only be regarded as reliable and acceptable when they have been collected by qualified entomological staff, or under their direct and competent supervision. The importance of supervision must be emphasized. In many cases the mass of data resulting from investigations prove to be of limited reliability and practical value owing to the lack of accurate collection of all the data necessary for the interpretation.

The entomological investigations are carried out in a limited number of representative areas, the number of which is adjusted in such a way as to allow adequate entomological observations to be carried out at the required interval by the entomologist, the assistant entomologist, or under their direct supervision.

In view of the difficulties in recruiting adequate entomological staff for national malaria programmes, it is necessary (when planning the entomological staffing pattern) not to budget for a complement of staff which simply cannot be selected, trained or recruited in the foreseeable future. At the same time, there are certain minimum requirements which it is desirable to retain and which, if necessary, may involve the recruitment or retention of international staff if the entomological services are to survive.

The possibility of recruiting and training national entomologists varies greatly from country to country, as does availability and quality. The requirements of the different countries may show wide differences according to size, nature of problem, phasing of malaria programme, etc. Consequently, it is difficult to make a rigid recommendation about the number and disposal of entomological staff applicable to all programmes. However, in order to offer some guidance some suggestions on staffing patterns are given below.

5.1 STAFFING REQUIREMENTS IN MALARIA CONTROL PROGRAMMES

In situations where significant expenditure on malaria control programmes is intended, the minimum basic entomological nucleus is considered to be:

- 1 fully trained entomologist;
- 2 high grade technicians, one for laboratory work and one for the field, and preferably interchangeable;
- 4 entomological aides, 2 assigned to each technician;
- other auxiliary personnel, e.g., collectors to be recruited locally according to needs, bearing in mind that bait capture may be made with local untrained volunteers.

With this kind of nucleus in a homogeneous area of 250 000 people, where residual insecticide is to be applied, a range of 4-6 indicator villages should be provided (one being in a comparison unsprayed area), located in most highly endemic parts of the area. Within each village, 4-8 baits should be employed, the distribution indoors and outdoors depending on the vector population involved. Five window traps would also be required.

Observations are expected to be made once a month. If the area is extended into areas of similar ecology and endemicity, then the indicator villages chosen should be redistributed in order to obtain optimum coverage of the larger area.

5.2 STAFFING REQUIREMENTS IN A MALARIA ERADICATION PROGRAMME

An example is given of a staffing pattern in a rather difficult tropical country with a population of approximately 20 million, in which there were three or four distinct endemic areas, widely separated and each requiring special entomological investigation. As this country, divided into six administrative regions, was already under total spray coverage, the entomological activities fell into three main categories:

- evaluation of spraying operations
- special observations in persisting foci of malaria transmission
- vigilance in consolidation areas.

In order to deal with these varied activities, the minimum requirement of entomological staff was considered to be:

- 1 senior entomologist, based in NMES headquarters (advised by WHO entomologist);
- 6 entomologists (regional), 1 based in each region;
- 5 senior technicians, 1 based in NMES headquarters, 4 based in the regions under attack;
- 1 technician (optional) based in NMES headquarters;
- 22 permanent insect collectors.

Of the 6 entomologists, 4 were based in different regions under attack for the evaluation of spraying operations; 1 was based in headquarters in charge of a mobile assessment team, and 1 was based in the regions in charge of a similar assessment team.

Of the 5 senior technicians, 4 were based in different regions for the evaluation of spraying operations, while 1 was based in NMES headquarters from which he made a regular itinerary of the consolidation areas.

The general disposal of entomological staff is shown in Table 6, further reference to the various duties of different staff being given in Schedules A, B, C and D.

With this disposal of entomological staff, routine observations are confined to 16 observation or index localities in the evaluation of spraying operations, and 8 localities with persisting foci in the more limited consolidation area. The 16 localities in sprayed areas were carefully selected on the basis of joint consideration by the senior malariologist and senior entomologist and the entomologist based in that particular region. These selected stations are intended to represent a wide range of conditions with regard to vector incidence and dominance, and with regard to different degrees of malaria endemicity.

The routine observations of both entomologists and senior technicians with regard to these permanent representative stations were limited to a series of straightforward capture or sampling methods, adequate to provide a general picture of vector incidence and reactions to insecticide treatment. Regular supervision of senior technicians by the entomologist in the regions was intended to ensure reasonable accuracy of observations and to assist in the early location or definition of difficult areas. Each locality is visited twice a month for 2-3 days, 8 stations being dealt with by the 4 entomologists and the remaining 8 by the 4 senior technicians.

For vigilance in the consolidation areas, a total of 8 stations were established among the localities with the highest malaria incidence in the past. Each station is visited by 1 senior technician and 2 insect collectors once every month during the known main transmission season for 1-2 days, this senior technician being based in NMES headquarters, from which he travels on a fixed circuit.

SCHEDULE A

Duties of senior entomologist

(WHO entomologist to assist in all phases of activities)

NMES headquarters

- Close liaison with the senior malariologist whom he assists in planning antivector measures.

- Supervision and participation in field work.

- Training of staff of various categories (basic and refresher training).

- Consolidation of field reports.

- Assessment of the entomological data in order to establish their epidemiological significance.

- Study of surveillance and special reports.

- Special entomological investigations anywhere in the country in the persisting foci of malaria transmission.

- Preparation of vector distribution maps.

- Preparation of vector incidence charts, month by month.

TABLE 6. SUMMARY OF ENTOMOLOGICAL ACTIVITIES OF THE VARIOUS CATEGORIES OF ENTOMOLOGICAL STAFF [a]

Activities	Staff
1. Evaluation in sprayed areas	4 senior technicians and 8 insect collectors to work in 8 stations, 2-3 days, twice a month
2. Special observations (assessment) in the proved foci of malaria transmission	2 entomologists and 4 insect collectors to work in areas wherever needed
3. Vigilance in consolidation areas	1 senior technician and 2 insect collectors to visit 8 selected stations, 1-2 days, once a month
	4 entomologists and 8 insect collectors to visit 8 stations 2-3 days, twice a month

[a] Proposed for a rather difficult tropical country with a population of about 20 million.

- Supplies and equipment for field teams; requirements to be submitted in advance and arrangements to be made for delivery of supplies and equipment to the field teams.

- Manual of operations for entomological work.

- Maintenance of live material in headquarters and establishment of mosquito colonies when needed and feasible.

- Preparation and maintenance of a collection of teaching material.

- Welfare of field entomological staff - adequate transport, payment of allowances, etc.

SCHEDULE B

Duties of assessment team entomologists

(WHO entomologist and the senior entomologist to assist and advise. Senior entomologist to supervise).

- NMES headquarters

- regional headquarters

These two entomologists will work in areas where foci of infection have been clearly defined and well established on the basis of surveillance work. Liaison with the surveillance teams is essential and suitable areas for work will have to be selected under the guidance of the team leaders where recurrence of transmission is suspected.

The assessment teams will select techniques according to the problem existing in the area.

SCHEDULE C

Duties of entomologists (regional)

(Supervised by senior entomologist. WHO entomologist to assist and advise)

Evaluation in sprayed areas

- Regular observations in two fixed localities. Each to be visited on consecutive days twice a month.

- Supervision of the assistant entomologist located in the area and arranging visits in such a way that between him and/or the senior entomologist at least one supervisory (working) visit is made each month.

- Assistance to the assessment entomologist on special observations.

- Liaison with surveillance workers.

SCHEDULE D

Duties of senior technicians

(senior entomologist and the entomologist (regional or assessment team) located in the area to supervise)

Regions I, II, V and VI

NMES headquarters (vigilance)

Evaluation in sprayed areas

- Regular observations in two localities (as for regional entomologist). Each locality to be visited for two days twice a month.

- Regular observations in localities with previous high receptivity mainly during the last part of the interval between spraying rounds and during the previously known period of peak density.

Vigilance in consolidation areas

The senior technician located in NMES headquarters will deal with this. Density studies carried out for two or three days only during the known peak(s) of vector density in localities known to have a high receptivity, and high vulnerability. The breeding potentialities will be estimated during these investigations.

6. ENTOMOLOGICAL ACTIVITIES FOR THE PLANNING AND EVALUATION OF ANTI-LARVAL OPERATIONS

6.1 INTRODUCTION

The entomological component is one of the essential elements for planning and evaluating anti-larval operations. It comprises several activities, the aim of which is to contribute to the assessment of the efficiency of the anti-larval operations. The entomological findings together with those of parasitological observations will determine the effectiveness of the operation and its impact on malaria transmission.

The attack on the larvae can be achieved by a variety of measures, such as source reduction, biological or chemical control. The entomological activities outlined here are basically meant for evaluating chemical larviciding but they could be adapted to evaluate some of the other methods of control (see section 4 of this Chapter).

Although anti-larval operations have always been included in urban malaria control, search of the literature shows that the subject of entomological evaluation has rarely been described in detail except for species eradication campaigns. Evaluation methods of such campaigns cannot be applied in an anti-larval programme that aims at reducing the vector density only to an epidemiologically insignificant level. On the other hand, there are a number of entomological evaluation methods which can be considered suitable for research studies.

In this section, the proposed procedures represent an attempt to lay down certain principles and to outline the techniques involved in the evaluation of anti-larval programmes. Flexibility in adaptation of the proposed procedures to local conditions should be exercised. As more experience in the entomological evaluation of anti-larval operations is gained, it is expected that planning and implementation of entomological activities can be made with more precision (for methods and techniques see Part II, section 7).

6.2 AN OUTLINE OF THE ENTOMOLOGICAL ACTIVITIES: FOR PLANNING AND EVALUATING ANTI-LARVAL OPERATIONS

For convenience in outlining the entomological activities, it is assumed that a programme is starting in an area which had no previous experience with anti-larval operations, so that the sequence of events may be followed as the programme develops. If it is decided that, for malaria control purposes, anti-larval measures may need to be implemented, it is essential that entomological element is involved from the beginning so as to contribute to the planning of the programme. As illustrated in Table 7, during the pre-treatment period the entomological

- 84 -

TABLE 7. THE ENTOMOLOGICAL ACTIVITIES IN AN ANTI-LARVAL PROGRAMME

Period	Activity		
Pre-treatment period (one year or more)	Preliminary survey	Entomological evaluation of larviciding trial	Trend observations (baseline)
First year of operation	Trend observations (trial area to be included)		Spot-checks
Second year	Trend observations		Spot-checks
Third and subsequent years	Trend observations*		Spot-checks

* To be terminated, or reduced to a minimum, if the efficacy of the dosage and frequency of application of the larvicide is confirmed; to be re-instituted and established in sections of the area where no response or a poor response has been consistently observed.

background information should be compiled and, where necessary, supplemented by observations so that it can be utilized in the initial planning of anti-larval operations. Having completed this, the entomological activities should be directed to the collection of baseline data by regular sampling of larval and adult densities in fixed capture stations. This information will show the seasonal trend of vector densities. During this pre-treatment period, there should also be a small-scale trial for determining the dosage and frequency of application of the larvicide proposed for use.

On completion of the collection of baseline data and with the commencement of the application of the larvicide, the entomological activities should be directed to the evaluation of its effectiveness and the adequacy of coverage.

The evaluation of the effectiveness of the larvicide depends primarily on continuing, for a certain period, observations in the same fixed capture stations which were established during the pre-treatment period to indicate the trend and to confirm the results obtained. Spot-checks should also be organized and extended to cover the area under treatment while the trend observations are gradually reduced.

However, trend observations may also need to be reimplemented in fixed stations to recheck the effectiveness of the larvicide in certain types of breeding place, which have not shown a favourable response, in order to assess, for example, the effect of an increased dosage or frequency.

The scope and scale of the entomological activities and the procedures to be followed are outlined below.

6.2.1 Pre-treatment activities

6.2.1.1 Preliminary surveys

Objectives

The main aim of these surveys is to provide background information which is a pre-requisite for planning of subsequent entomological activities, and for the initial planning of anti-larval operations.

Timing and duration

The preliminary survey should be carried out during the seasons of vector(s) prevalence. It should be undertaken on a sample basis representing different ecological conditions in the area under reference. Depending on the extent of the area and the availability of staff and facilities, the preliminary survey may be completed in 1-3 months.

Procedures

The first step is to collect from the existing documents, reports and literature, relevant information on:

- the environmental conditions including seasonal meteorological changes;

- the anopheline fauna with special reference to recognized malaria vectors and suspected vectors;

- the ecology of vectors and suspected vectors, including their breeding habits, duration of the period of the immature stages, adult flight range, etc.;

- the season of malaria transmission.

The second step is to carry out adult and larval surveys:

- to check whether the recorded information on the proved and suspected vectors of anophelines is reliable or needs to be supplemented.

- to identify the breeding places of the vectors, their nature and approximately their extent, and to classify them into different types according to the physical conditions and the presence of salinity and pollution (see Chapter 1).

- to collect information on the suitability of some larval breeding places and adult day-time resting shelters for use as fixed capture stations.

6.2.1.2 Baseline trend observations

Objectives

To establish baseline data on densities of larvae and adults and on the susceptibility levels to insecticides of malaria vectors and suspected vectors throughout the different seasons; and to study the influence of environmental conditions on larval and adult output.

During this period of observation, a small-scale trial should be organized for determining the dosage and frequency of application of the larvicide to be used. In this trial, regular observations on larvae should be conducted as described below. The results of trend observations and the trial will assist in adjusting the plan of anti-larval operations.

Timing and duration

The regular observations on adult vectors should be started shortly after the completion of the preliminary survey. They should continue for a complete year before commencing the anti-larval operations.

Procedures

The trend observations should be carried out on both larval and adult populations in fixed indicator or follow-up units[1] to be selected to represent each ecological substratum in the area designated for anti-larval operations.

A. Procedures for larval observations

(i) Establishment of indicator units and fixed capture stations

From the information gathered during the preliminary surveys, a number of these indicator units should be selected. In each a number of breeding places representing the breeding places of malaria vectors should be selected on the basis of high larval density. These selected breeding places should be established as fixed capture stations for regular checking. In each fixed capture station, a standard number of dips should be made at intervals of 7-10 days or longer, depending on the life-span of the aquatic stages in different seasons.

(ii) Establishing baseline data for the susceptibility levels of vectors

Susceptibility tests must be carried out during each season on larvae of malaria vectors with the insecticide used in larviciding as well as with other insecticides for which test kits are available. (If impregnated papers are available for the same insecticide, adult susceptibility tests should also be carried out.)

---

[1] An indicator unit is to be established in an operational section in the area planned to be allocated to a larviciding squad according to geographical reconnaissance.

B. **Procedures for adult observations**

(i) *House-resting density*

In each indicator area, at least 10 adult capture stations having high vector density should be selected in the vicinity of larval capture stations. Collection of adults should preferably be carried out by hand/spray capture every two weeks.

(ii) *Vector/man contact*

In a selected number of adult capture stations, bait capture on man indoors and outdoors should be carried out at fortnightly intervals in order to establish the baseline for vector biting density and its seasonal trends.

(iii) *Light-traps*

These devices can provide an additional tool for studying the changes in the vector density. It is an advantage if the light-traps have been operated during the pre-treatment period as this provides experience on the best location and operation (including the frequency of captures) of this type of mechanical device.

6.2.2 Entomological evaluation of anti-larval operations

6.2.2.1 General objectives

The general objective is to assess the efficiency of the anti-larval operations in reducing vector density. The entomological parameters together with the parasitological findings will enable an assessment to be made of the impact of the operation on malaria transmission. The specific objectives are given under each procedure.

6.2.2.2 Trend observations

Timing and duration

When the anti-larval operations commence after the completion of baseline observations, the trend observations should continue in the same fixed capture stations, at the same indicator units selected prior to treatment. The larval and adult checking should continue in the first year of operations regularly throughout the operational season and, as mentioned earlier, may need to be continued at longer intervals during the off-season. Trend observations may extend to the second year to confirm the findings of the first year. Thereafter, to assess the continued efficacy of remedial measures, they should be maintained in a limited number of units only and in areas where unfavourable response was encountered. Naturally, trend observations should be reactivated whenever the larvicide has to be replaced by another.

Procedures

A. **Procedures for larval observations of vectors**

The specific aim here is to maintain a continuous check on the efficacy of the designed dosage and frequency of application on the density of the vector population under different ecological conditions. For this, the trend observations should be continued at the same fixed larval capture stations, using the same standardized techniques of sampling. For a description of larval sampling techniques see Part II, Section 7.2. With this specific aim, those fixed larval stations, and the area they represent, should receive the planned treatment well applied. There is a tendency of larviciding teams to overdose the fixed capture stations. It is, therefore, necessary that these capture stations and other breeding places and indicator areas be larvicided under close supervision to ensure that the dosage is applied at the appropriate time. Otherwise, the results of the trend observations may suffer as a result of the variability of application and hence lead to confusing results because of the operational deficiencies. The timing of the larval checking should coincide with the last day of the larviciding interval which had been determined by the initial trial. The presence of the third and fourth stage larvae of vector anophelines or their pupae should be recorded and the reasons for such failure should be investigated.

B. Procedures for adult observations

The specific objective is to determine the effect of anti-larval measures on adult densities of vectors. Larviciding essentially operates on the larval density, thereby reducing the adult output. The trend observations to be carried out on the adult vector population in the indicator units should reveal whether the larvicide as applied under optimum conditions of adequate dosage, frequency and coverage has succeeded in reducing the adult density to an epidemiologically insignificant level.

House-resting density. House searches should be continued at the same fixed capture stations from which the baseline data were collected, using the same technique.

Man/vector contact. Bait capture should continue in the same capture stations using the same number of baits. Both house-resting and bait capture should be maintained at fortnightly intervals.

Light-traps. If light-traps have been found to be suitable for sampling, a number of them should be operated in each indicator unit at fortnightly intervals.

6.2.2.3 Spot-checks

Objectives

As has already been mentioned, the trend observations at fixed indicator units would reflect the effectiveness of the larvicide as adequately applied at the designed dosage and frequency of application. Spot-checks, on the other hand, should aim at detection of operational deficiencies which may occur in the large part of the treated area, not within indicator units, so that remedial measures can be implemented in time. In addition, spot-checks should aim at revealing failure of the larvicide to cope with abnormal physical and environmental changes which may occur in certain areas in certain seasons.

Timing and duration

Spot-checks should commence simultaneously with trend observations as soon as the anti-larval operations have started. The number of indicator units is reduced in the second and subsequent years to a minimum which would allow density trends to be verified. The time saved may be used to increase spot-check activities which should continue as long as the larval control operation continues.

Procedures

In spot-checks besides inspection of breeding places, adult searches should be undertaken. Adult searches should be made by hand/spray capture in about 10 premises in each area selected for spot-checking. Light-traps may be tried as an additional method. In larval searches, the standardized units of 10 dips should be adopted at each breeding place selected for inspection. However, there may be a need to increase the units of dips by twofold or more, for example, when breeding is suspected on account of finding a few early stage larvae or adult mosquitos in premises. If the breeding place proves to be negative, the presence of adults would indicate the necessity of extending the spot checks to adjacent areas.

Ideally, all breeding places should be inspected by spot-checks but this would not be feasible in a large-scale larval control programme. Therefore, spot-checks can be implemented on a sampling basis. For example, if the area of operation is divided into four sections, inspection of 10-12% of the breeding places of each section each week on a rotational basis by the entomology teams would permit an assessment of the quality of larviciding in about 40-50% of the breeding places in the whole operational area once every month. Besides which, there should be routine inspection for larvae undertaken by the operational supervisors in the normal course of their duty. Both types of inspections should be coordinated so that they cover as many of the breeding places as possible in each of the operational units of each larvicider. In the selection of areas for spot-checks, priority should be given to those which are suspected on an epidemiological basis where searches should, therefore, be intensified. The entomological teams carrying out spot-checks should be supplied with copies of the schedules of application of the larvicide in each operational unit. Searches for adult mosquitos should be made around all breeding places that have been found positive.

6.2.3 Recording, summarizing and interpreting data

6.2.3.1 Information to be recorded

In preliminary surveys

The following information needs to be collected:

A. Description of the area

The area should be divided tentatively into ecological substrata, to be elaborated later when geographical reconnaissance of breeding places is undertaken.

B. Meteorological conditions

Meteorological records should be obtained from the existing meteorological stations for the whole area for the past ten years if possible. Specific information on conditions affecting the breeding places at the time of the survey should be collected.

C. Description of the breeding place and physical conditions

- Type, and whether water is permanent or transient.

- An approximate estimation of the surface area and the depth of the water in each type of breeding place.

- The presence and types of vegetation and algae, etc., and degree of their growth.

- The presence of current and shade.

- The presence of salinity or pollution.

D. The number of dips (a minimum of 10 dips or multiples of 10)

Since in this initial survey, it is not possible to standardize the number of units of dips at each breeding place, the exact number of units of 10 dips and the number of positive units of 10 dips should be recorded. A note should be made regarding which breeding places might be suitable as fixed capture stations.

E. Description of the different types of day-time shelters surveyed for adult vectors

A note should be made on those found suitable as fixed capture stations.

In trend observations

At fixed capture stations, the above information (A and C) should be collected during the initial observations. Subsequently, only the important changes in the condition of the breeding places should be recorded. The number of units of 10 dips (or multiples of 10) should be standardized at each fixed capture station selected for trend observations and maintained as such throughout the period of observations.

Similarly, the day-time resting shelters selected as fixed capture stations should be described at the outset. Meteorological data should be collected regularly.

In spot-checks

Since this is a rapid survey to detect whether the breeding place is positive or negative for larvae and pupae, records of the type of each breeding place and its environmental conditions should be made only when it is found positive. The dates of application of the larvicide must be mentioned (these are to be obtained from the larviciding operational schedules).

6.2.3.2 Summarizing the data

(Data of trend observations and spot-checks should be summarized separately).

Larval density indices of vectors

(a) Samples should be classified into:

- the group of 1st-2nd instar larvae

- the group of 3rd and 4th instar larvae for each vector species collected

- other anopheline larvae (pooled)

- anopheline pupae

(b) The index of larval density of 3rd-4th instar larvae should be calculated to give the average number per unit of 10 (or multiples of 10) dips:

- for positive capture stations of each type of breeding place inspected,

- for positive capture stations of all types of breeding places inspected,

- for all capture stations inspected of each type of breeding place inspected,

- for all capture stations inspected of all types of breeding places inspected.

In addition to the above, it would be useful for presentation of the results and assessment of the effectiveness of the operations to establish the frequency distribution of the breeding places according to their larval density using the appropriate class intervals. This would show the number of breeding places which have given high larval density to which special attention should be given for investigation to be carried out and remedial action and follow-up to be applied. Larval indices should be worked out separately for samples obtained by different devices (see Part II, section 7), for example:

- index for ladle collections

- net dipper in surface waters

- net dipper in wells.

Adult density indices of vectors

The usual average number of vector females per room in day-time searches and average bites of vectors per man per night in bait capture should be worked out.

6.2.3.3 Interpretation of results

It is essential that the interpretation of the results obtained from the above-mentioned activities during the post-treatment period be made progressively in order that timely remedial action can be implemented.

For progressive assessment, the entomological parameters should be presented graphically, together with the parasitological and meteorological data as soon as the data is collected, i.e., each week for larvae and every fortnight for adults.

Providing that there were no operational defects, the consistent presence, shown through the larval indices obtained from the fixed capture station, of advanced instar larvae or pupae of anopheline vectors at the end of the interval of application would indicate that the larvicide applied at the optimum field dosage and frequency was not effective, or that the susceptibility of the Anopheles vectors to the larvicide has deteriorated. This will call for further close observations of the affected breeding places involving also assessment of the prevailing environmental conditions, as temporary physical changes may have been responsible. In the light of the results of such investigation and results of adult capture as well as results of spot-checks, a revision of the rate and frequency of application may be required in areas affected (or after due consideration of all factors, in the whole area of operations).

The adult house-resting densities obtained from the indicator units only reflect the impact of the larviciding applied in those areas, if the species rests mainly indoors. However, with a vector exhibiting exophilic tendencies, it is virtually impossible to determine the proportions that rest indoors and outdoors, and the reduction in density of the outside resting population therefore is extremely difficult to estimate. Hence, the biting indices, particularly when the species involved has exophilic tendency, also help to determine the net impact on the man/vector contact and whether it is still of epidemiological significance.

A knowledge of the critical density of vector in relation to man (which varies according to the efficiency of the vector species) is an asset, as it facilitates the interpretation of levels of biting densities. For this reason, a close liaison between the entomologist and the malariologist is a pre-requisite for determining by joint analysis of entomological/parasitological data, the level of critical density, if this cannot be easily ascertained from existing information (see Part I, Chapter 5).

The light-traps may reveal the presence of the vector, when the other methods fail to do so.

However, the adult parameters obtained from regularly observed indicator units may also reflect operational deficiencies occurring in the surrounding area. To eliminate this possibility, the surrounding areas should be subjected to spot-checks.

In spot-checks, the presence of the advanced instar larvae and pupae of anopheline species in the breeding places inspected should immediately be investigated in order to establish whether this was due to negligence on the part of the larvicider or to the inefficiency of the larvicide to cope with a particular environmental situation. Timely remedial action can then be applied as appropriate.

6.2.4 Organizational aspects

6.2.4.1 Staffing

The staff required for carrying out the entomological activities in a larviciding programme will depend on the following:

- the size of the area of operations;

- the number of breeding places, their surface area and accessibility;

- the distribution of the breeding places with respect to the distance to be travelled by the insect collector.

For this, experience of the preliminary surveys will make it possible to estimate:

- the average number of breeding places that can be inspected by a collector per day in each section of the operational area; and

- the number of collectors needed for carrying out larval checks and adult mosquito captures.

The following example may illustrate the staffing needs in an area having breeding places reasonably distributed, where water is accessible and the larval density is moderate.

One collector should be able to cover 12-18 larval capture stations per 6-hour day, assuming that about half of that time would be spent walking from one station to another.

If in an area there are 100-150 fixed larval capture stations to be observed at weekly intervals in 4 indicator units to be checked for adult vectors with bait and spray capture every two weeks, a team of 8 insect collectors would be able to undertake the following weekly schedule:

- cover all larval fixed capture stations in one day;

- carry out 2 night observations in two of the four indicator units (alternating with the other two units every second week);

- carry out spray capture for adult mosquitos in the morning following night capture (or, if more convenient, on a separate visit).

To such a team, one supervisor and one technician should be assigned. Both will conduct the periodical susceptibility testing, operate the light-traps and summarize the reports of the collectors. The insect collectors may be trained to identify microscopically the larvae and adults (if their educational standards permit) under the supervision of the technician and the supervisor.

After the treatment has started, additional personnel would be needed to carry out spot-checks. If the above area had about 1000-1500 breeding places, a team of 4 insect collectors would be able to inspect about 20-25% of the breeding places each week. This team would need one supervisor. A senior technician should be assigned for the overall supervision and direction of the entomological activities under the guidance of the senior entomologist of the malaria control section in the public health service.

6.2.4.2 Transport

Adequate transport of the appropriate type should be provided for the teams and the supervisors depending on the extent of the area under operation.

6.2.4.3   Supplies and equipment

   - kit for collecting larvae,

   - kit for collecting adults,

   - WHO standard susceptibility test kit for larvae,

   - WHO standard susceptibility test kit for adults,

   - laboratory equipment and supplies for examination of larvae and adults.

6.3   ENTOMOLOGICAL EVALUATION OF TRIALS WITH CHEMICAL LARVICIDES

6.3.1   Objectives

An essential element in planning a larviciding programme, particularly when it is intended to use a new larvicide, is that information on the appropriate formulation, dosage and frequency of application should be obtained from a small-scale trial carried out locally by: (a) selecting the suitable formulation(s) for the prevailing types of breeding places; and (b) determining the dosage and frequency of application of the formulation(s) selected throughout the different seasons.

6.3.2   Timing and duration

The trial must be carried out as soon as possible after the decision has been made to apply larviciding measures. It should be carried out in the preparatory year while collecting all the necessary baseline data and other information required for planning. However, it must be carried out during the favourable breeding season of vectors and should satisfy the investigator of the adequacy of the information on the formulation, dosage and frequency of application. A period of three months should be sufficient during each of the wet and dry seasons.

6.3.3   Procedures

Checking of the adult density cannot be considered reliable in a limited size trial because of the infiltration of mosquitos from the surrounding untreated area. Therefore, the evaluation will rely on larval checking.

6.3.3.1   Pre-treatment larval survey

In the area chosen for undertaking the trial, a number of breeding places should be selected as fixed larval capture stations in which larval checking should be maintained using a standard number of dips (see Annex 1). Larval sampling should be made in order to establish baseline data for at least a month before application of the larvicide during the favourable season. An area having similar types of breeding places should be selected to serve as a comparison area in which a similar number of standardized dips are to be made in fixed capture stations.

6.3.3.2   Post-treatment observations

After the application of the larvicide, the same standard number of dips should be made in each of the fixed larval capture stations. The post-treatment observations should be made as follows:

(a) one day after treatment in a section of each type of breeding place. This will provide information on the immediate effect of the larvicide.

(b) on the last day of the interval between applications as indicated from available information on larval life span and residual effectiveness of the larvicide, in another section of the breeding places;

(c) the same observations should be maintained in the comparison area.

6.3.4 Recording, summarizing and interpreting the data

Section 6.2.3.3 of this Chapter is concerned with interpretation of larval indices. The results of the trial, together with those of baseline observations, will indicate the final adjustments required in the plan of anti-larval measures.

BIBLIOGRAPHY

WHO Expert Committee on Malaria (1962) Ninth Report, Geneva, Wld Hlth Org. techn. Rep. Ser., No. 243

WHO Expert Committee on Malaria (1964) Eleventh Report, Geneva, Wld Hlth Org. techn. Rep. Ser., No. 291

WHO Expert Committee on Malaria (1973) Sixteenth Report, Geneva, Wld Hlth Org. techn. Rep. Ser., No. 549

World Health Organization (1963) Terminology of malaria and malaria eradication, Geneva

Malaria control in countries where time-limited eradication is impracticable at present (1974) Wld Hlth Org. techn. Rep. Ser., No. 537

Manual on Larval Control Operations in Malaria Programmes (1973) World Health Organization, Geneva

ANNEX 1

THE SELECTION OF INDICATOR VILLAGES AND CAPTURE STATIONS

In the course of the preliminary survey, the entomologist should become familiar with the region in which he is going to work, with the topography, distribution and nature of villages or human settlements, with the communications and accessibility of different areas, water sources and various breeding places. All this preliminary exploration or reconnaissance will form a sound basis for later more systematic and extended survey, and for an ultimate zone-wide, or country-wide, system of indicator villages or capture stations.

In the selection of indicator villages and capture stations there are two main objectives; firstly, the selection of a suitable locality in which the density of vectors will permit necessary studies on vector habits, resting, biting habits, distribution in indoor and outdoor shelters, and - later - the reactions of vectors to insecticide treatment in local and experimental huts or houses. The second objective is to organize a system of representative collecting stations or observation villages in which a routine of entomological work - established before spraying begins - is best adapted for subsequent overall evaluation of spraying operations in different zones or country as a whole.

While purely entomological considerations may decide the selection of study localities for the first objective, in the case of the second objective, selection must be governed primarily by malaria endemicity. With that in view, therefore, the selection of such villages and capture stations will be guided by the following considerations.

Malaria endemicity

Observation villages should be selected jointly by the national/regional entomologist and the malariologist according to areas of high, medium or low endemicity, and according to different environmental conditions. The selection of epidemiologically homogeneous areas representative of the country as a whole should be determined on the basis of a careful study of the available malariometric data and on the basis of geographical and entomological data. The observations will be mainly carried out in localities with high endemicity. Regular malariometric surveys, while planned on a wider sampling basis than the entomological work, should invariably include child and infant parasite rates in the indicator villages. Quarterly data on the child and infant parasite rates in these villages are desirable through the preparatory and early-attack phases.

Topography

Observation villages should also be selected in such a way as to represent the main ecological types in the project area, e.g., forests, foothill areas, cultivated plains, river valleys, irrigation areas, etc. There is often a close general agreement between the topography, vector distribution, and breeding potentialities and the malaria endemicity. The selection of localities may, by taking into account the topographical and ecological aspect, reveal differences in seasonal incidence, differences in the transmission season, and even differences in the identity of the dominant vector within the same uniform endemic zone. Such differences may well play an important part in relation to the timing of spraying operations, and to the early anticipation of the existence of difficult or problem areas.

Selection according to topography is of particular importance in foothill and mountainous regions where it is important to have an accurate idea - based on joint consideration of entomology, malariometric data and mean daily temperature records, of the altitude levels above which transmission becomes of negligible importance.

Annex 1

Areas of vector dominance

If malaria in the project area is transmitted by more than one main vector, indicator villages representative of areas of dominance of each of the main vectors should be selected. The transition between different areas of dominance may well be determined by geographical features, and in certain types of terrain, for example in coastal areas, there may be a sharp transition from one dominant vector to another. Allowance must be made for this in the selection or grouping of capture stations. The extent of the favourable environment for adults and of the characteristic breeding places are the most important elements in determining the dominance of a species. Man may produce important modifications in the dominance of vectors by modifying the normal habitat by cutting the jungle, development of irrigation, etc.

Areas of different seasonal incidence of the vector

As the output of vectors is determined by the range and suitability of favourable breeding places - apart from climatic factors - it follows that the distribution and abundance of vectors within an area will change according to seasonal changes in the breeding foci. For example, breeding places may be widely dispersed in the rainy season but strictly limited to perennial streams or swamps in the dry season. In the planning of indicator villages and capture stations this should be taken into account so that an adequate overall picture is presented of vector seasonal incidence related to the type and variability of breeding potentialities. Indicator villages selected in the rainy season, for example, may produce such low collections in the dry weather giving the impression of general vector scarcity and may in this way completely fail to detect the persistence of adult vectors in certain dry weather foci. When anti-vector measures are applied, the evaluation of the impact of such measures on local vectors is important during the main, favourable, development period, but it should be borne in mind that the species will usually be most vulnerable to attack during a phase of contracted breeding opportunity. As an example, in savannah areas, A. gambiae is a predominant species during the rains whereas A. funestus predominates during the dry season.

Communications, and accessibility of indicator villages

As the value of entomological data obtained in selected collecting stations or observation sites is greatly enhanced if records are continued regularly throughout all the period of vector activity, it follows that in the final choice of villages representing particular endemic or topographic conditions, the question of accessibility at all seasons is important. Difficulty of access due to floods or impassable roads may have to be reckoned with at intermittent periods in the rainy season and will not necessarily disrupt the regularity of records. But if communications are such that an observation area is isolated or inaccessible for two to three months of the year during the active period of development of vectors, its value will be considerably reduced.

Number of observation villages

In planning zone or country-wide observation villages on the representative principles mentioned above, the entomologist faced with limitations of staff, transport and time has to decide the best course of action to achieve the objectives: whether to deal with a large number of observation or index villages representing administrative divisions as well, such as sector, unit, or sub-unit - which can only be dealt with briefly at longer intervals, or whether to restrict himself to a comparatively limited number of representative localities for each type of homogeneous ecological area which can be dealt with much more fully and regularly.

The only entomological data likely to be of real value are those which have been collected by qualified entomologists or assistant entomologists, or under the supervision of such qualified staff. The practice of leaving the collection of data entirely to subordinate staff should not be encouraged, even though it may appear that the use of such staff is unavoidable

Annex 1

if a wide representative coverage is to be aimed at. Entomological survey and evaluation in malaria eradication programmes is a specialized activity which must be carried out and supervised by qualified staff.

In view of the wide variation of conditions occurring in various countries, as regards availability of staff, transport, local customs, budget, etc., it is difficult to lay down rigid rules about the actual number and disposition of indicator villages, the actual number of collecting stations in each village, or the frequency of visits. In principle, two villages for a representative epidemiological area are the absolute minimum for follow-up collection; four are preferable. The most important representative villages should have records taken by fortnightly visits. A distant observation village may be dealt with adequately by a two to three day visit once a month during which the basic sampling methods which have been selected are operated. These include resting densities in sprayed and unsprayed houses (hand-collection or space spray as appropriate), night collections with human bait indoors and outdoors, collections in artificial outdoor shelters and window trap collection observations where possible. Indicator villages within two to three hours' drive from the field laboratory may be visited for shorter periods more frequently, live mosquito material being preferably brought back to the field laboratory for examination. In this way the field laboratory may form the centre of an area sufficiently large - perhaps up to 160 km in diameter - to include the main representative indicator villages. Guidance about the number of indicator villages and a <u>minimum</u> of staff is given in Chapter 2, Section 5.

<u>Arrangement of collecting stations in observation villages</u>

Many entomologists and their staff of mosquito collectors concentrate on those houses or habitations where they can usually rely on finding the greatest number of vectors in the shortest time. When collections in these attractive houses decrease to a low level, it is assumed that this indicates a corresponding decrease in the vector population as a whole; in order to confirm this random collections are also necessary. When studying the daily and seasonal fluctuations, and considering the to-and-fro movements of vector populations in a locality, the actual distribution of capture stations within the observation village should include all types of house at various distances from the breeding places.

When monitoring the impact of spraying on a mosquito population the most attractive houses should be selected for regular observations; in addition extra sampling should be carried out as much as possible, in those areas of the villages and types of houses which normally might have the highest density in the investigated locality. Spot checks are of considerable importance for completing the information on vector density, and help in detecting operational failures or various modifications caused by environmental changes, as noted above.

Reference has already been made to the fact that the villages themselves should be selected in such a way as to deal with changing vector distribution at different seasons. In the same way, within the village itself, allowance must be made for the day-to-day or seasonal shift or movement of vector populations. Position and type of breeding place should be considered. (When studying the relative general density rather more random and unbiased distribution of capture stations within the village should be established.)

Fig. 18 below shows the suggested distribution of capture stations in relation to breeding places. Each capture station would be made up of two or three houses (huts) where indoor resting densities, and indoor and outdoor (peridomestic) human bait collection, etc. should be carried out and two or three artificial outdoor shelters (pit-trap, box-trap, etc.).[1]

---

[1] Regular collection in artificial shelters is not entirely necessary when studying the mosquito population by indoor resting densities and/or man-mosquito contact. This might be used in special studies, e.g., for precipitin tests or as an additional method for obtaining mosquitos. Such collections are not successful with all vector species.

Annex 1

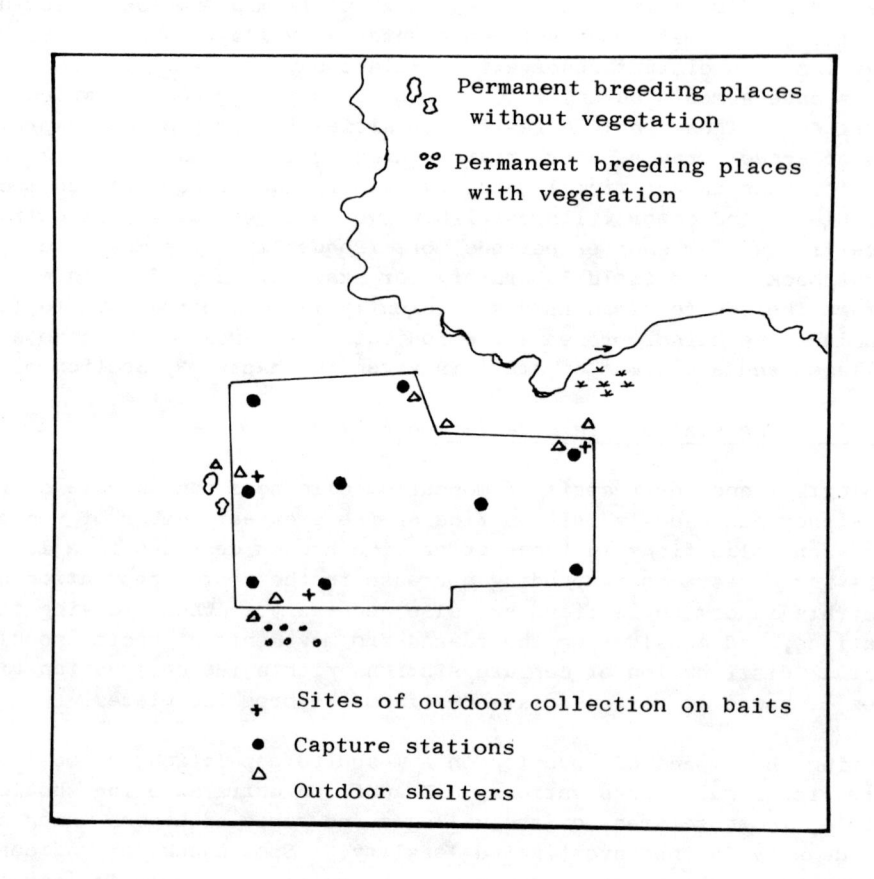

Fig. 18   Suggested distribution of capture stations
in and around observation villages

Annex 1

These samples would be supplemented from time to time by extradomestic - as distinct from peridomestic - captures on human or animal bait at points outside the premises, in the spots where people or animals might stay during the night (inside or outside the locality). Such observations are carried out only for a short time during the peak density period.

Depending on staff and time available, the routine observations could be carried out in all stations at each visit or by an alternating system of half of the stations at a time. The exact details of how this could be operated might be expected to vary according to the different countries, different vectors and objectives and would have to be decided by the entomologist on the spot. In practice, the selection and number of capture stations should make possible a reasonably reliable sampling in order to obtain representative data. When studying the impact of spraying on the vector relative density the investigations should be concentrated only in those fixed capture stations which had the highest productivity before spraying. Collection on human baits is an essential method for the evaluation of the impact of spraying on the vector population and transmission, and is the main method for assessing the impact of anti-vector measures on exophilic vectors. The number of places is limited to 2-3 in each village for indoor and outdoor collection. Therefore the selection of such places should be made on the basis of preliminary investigations in order to find the most productive capture stations.

CHAPTER 3.   CRITICAL APPRAISAL OF TECHNIQUES OF SAMPLING MOSQUITO POPULATIONS

CONTENTS

		Page
1.	INTRODUCTION	102
2.	REVIEW OF VARIOUS SAMPLING METHODS	103
	BIBLIOGRAPHY	108

CHAPTER 3

CRITICAL APPRAISAL OF TECHNIQUES OF SAMPLING MOSQUITO POPULATIONS

1. INTRODUCTION

As it is impossible to examine the entire population of mosquitos in different habitats spread over large areas, sampling is carried out.

A sample of mosquitos is a representative part of a mosquito population collected over a fixed period of time in a given situation within its habitat. Since a population of mosquitos is composed of individuals that possess the same basic characteristics which vary in different individuals between some known normal limits due to the physiological status and environmental factors, a sample of a certain number of individuals of a given population will be satisfactory for the study of the characteristics common to the whole mosquito population in that area, e.g., variability of the density, various aspects of behaviour, infectivity, etc.

The main objective of sampling techniques is to obtain the maximum accuracy of information for the time and money expended. According to statistical theory, the sample estimate should not lie outside the limits of plus or minus twice the standard error in more than 5% of cases. A smaller standard error requires a larger sample size. A decision to accept a larger standard error may diminish the value of the result, but a demand for unnecessarily high precision may mean a waste of resources.

In performing the sampling the following factors should be taken into account:

- The objectives of the investigation, and information needed.

- The species to be investigated.

- Environmental conditions in the sites selected for sampling, e.g., type and characteristics of the site, presence of hosts, winds, rain, phase of the moon, etc.

- Efficacy and limitations (feasibility) of the sampling method taking into account the information mentioned above.

- The application of the sampling method and techniques should be accurate and of satisfactory duration. The intervals between repeated collections of samples in the same spot should be such as to offer the possibility of recording the normal changes occurring in a given population of mosquitos.

- The sites investigated and the number of individuals sampled should be representative for relatively homogeneous areas.

Entomological sampling techniques have been giving satisfactory results in field projects where adequate staffing and supervision usually exist. Nevertheless, there are occasions where the validity of the entomological indices has been doubtful. Apart from the wide variation in the efficiency of the sampling techniques under the influence of environmental changes, the sensitivity of these methods is much affected when staffing and supervision is inadequate, and further complications arise with the intervention of logistic problems. However, the sampling techniques are particularly productive when applied in unsprayed areas or after discontinuation of the attack measures in areas where the vector density regains its original level. The sensitivity of sampling techniques is severely tested when dynamic changes take place in a vector population such as those which occur under the influence of attack measures. Discussion is confined to sampling techniques as used for evaluation of attack measures in a malaria eradication or malaria control programme.

A few villages should be selected among those which are highly malarious within a larger area homogeneous in respect of vector species. In order to assess the impact of residual house spraying without operational defects, the spraying of indicator villages should be made under close supervision to ensure that no operational deficiencies occur. The indicator villages can then be considered as a good guide to the level of control to be expected under optimum conditions and their importance in establishing baseline data and in longitudinal assessment of the programme is underlined.

A sample of this kind, however, must not be regarded as a valid statistical sample of the whole area, but, in a malaria programme the interest is primarily in ascertaining how the malaria situation is changing as the programme proceeds. The impact on the vector in areas typical of different epidemiological characteristics is the main entomological factor, and, provided the indicator villages have been properly selected in the first instance and operational quality is reasonably uniform, that is what will emerge from this type of sampling.

Normally the fixed indicator villages selected in a given area will include those where malaria is most highly endemic, thus while such a sample is admittedly biased, it is biased in the direction of the more difficult or refractory localities, hence favourable indications derived from samples from these localities may be extended to the rest of the area with a fair degree of confidence.

2. REVIEW OF VARIOUS SAMPLING METHODS

Sampling indoor resting density

The methods of estimation of indoor resting density which are commonly used in malaria programmes are hand capture and spray capture. The first method, despite its unreliability, has been used in situations where open types of houses exist and/or where obstructions such as household objects, furniture and grain stores occur. However, the second method is the most reliable when the condition of the premises is suitable. With endophilic vectors, and prior to the use of residual insecticides, spray capture is an objective method for it gives an indication of the biting activity of the previous night, and an average of 20-24 houses a day can be covered with 4-5 operators. The objectives of indoor resting surveys and the essential requirements for their utilization can be outlined as follows.

In regular observations. The indoor resting density will serve as a guide to the impact of the insecticide on this fraction of the endophilic vector population. No substitute methods for the conventional hand/spray capture could be suggested since this technique provides an estimate of the room density. The premises to be selected for indoor resting surveys of spray capture should be those giving high yields of the vector and not likely to be demolished or reconstructed. Their efficiency in respect of window trap catches should also be taken into consideration. Here also, it is stressed that in order to assess the impact of the insecticide house spraying without operational defects, the spraying of these capture stations should receive closer supervision to ensure that they are neither undersprayed nor oversprayed. However, the post-spraying indoor density estimates in fixed capture stations are of limited value, particularly where irritating insecticides are involved. Indicator villages should also be covered by spot-checks.

In spot-checks. In areas under attack, the indoor resting density surveys for an endophilic vector can be utilized as a means of detection of operational defects. In order to achieve a reasonable geographical coverage by spot-checks, a simple method of detection of house resting is considered. Partial spray-catch with high concentration (0.3%) pyrethrum solution and portable sheet is advocated (2-3 man team). This has been known as the "umbrella" method of spray catch (Logan, 1953).

This would take roughly half of the time required for the full spray-sheet method; thus more rooms can be surveyed. Since this method only samples part of the indoor resting population, the relation of its yield to that of the full spray-sheet method should be assessed. This will help in comparing with the results of the latter method in the fixed capture stations of the index localities. In surveying the localities, searches should be made for groups of missed houses and new structures and a sample of apparently sprayed houses should also be surveyed.

The timing of spray-catches is not of great importance in premises not sprayed with residual insecticides except that it is more convenient to do it early in the day to allow time for examination of material collected. Where the combination of a slow-acting insecticide and dawn-influx of mosquitos is known to occur as at Kisumu, Kenya and El Salvador where fenitrothion was being used, late-in-the-day spraying is essential. The presence of live mosquitos resting on sprayed surfaces late in the day suggests possible resistance by the vector to the insecticide, provided the houses are proved to have been properly sprayed and the observations are made within the period of the effectiveness of the insecticide. Susceptibility testing should be carried out to confirm the presence of insecticide resistance. Susceptibility testing should however be performed not only in situations where a high vector density is encountered in sprayed areas, but also as periodical checking for early detection of incipient resistance, in which case vector density may still be at a relatively low level.

The frequency and extent of such spot-checks must ultimately depend on the number of entomological staff available. The ideal would be to have an independent mobile entomological investigation team to perform such activities as well as other special investigations connected with persistence of transmission.

In order to achieve the main objectives of spot-checks, the results should be rapidly transmitted to the operational officer to enable him to take necessary remedial action where indicated.

Sampling for estimation of man-vector contact

The objective of these procedures is to determine the biting component of the mosquito population. This estimate has been termed the man-biting index (MBI). The commonly adopted method utilizing one or two baits indoors and outdoors does not give an index which can be used for making valid inferences on the general population. However, under unsprayed conditions when the density is high, the method may still remain above its lowest sensitivity level. Once spraying has taken place and density decreases, sensitivity falls and this same number of baits is unlikely to detect biting, hence the results are erratic and of questionable value. To illustrate this statistically, if in a village of 1000 inhabitants there are 200 mosquitos coming to bite, the chance of a man being bitten is 0.2, and, assuming a Poisson distribution of the bites, the probability that two men would receive no bites is 67 in 100.

It is suggested that several baits must be used; 16-20 baits divided into groups alternating in shifts of work and rest during the night. It was also generally agreed that to overcome staffing problems which would be associated with the provision of such large bait numbers, the substitution of some kind of human bait net trap using local volunteers should be developed.

Direct comparison of efficiency of all-night human bait catch and net trap catch needs to be made. Such net traps could be used for both indoor and outdoor catches. The type of net needs further study, particularly in relation to the entry of females which have fed elsewhere. It is also recommended that no restrictive partition preventing entered mosquitos from feeding on the enclosed bait should be incorporated. It must be remembered, however, that the human bait net trap will only record the number of mosquitos coming to bite during sleeping hours, unless special arrangements are made to ensure its occupation by the bait during all the hours of mosquito activity.

The man-biting index is considered by most entomologists to represent the most important factor in estimating the entomological inoculation rate. In most cases where investigations have been made, the entomological inoculation rate has been much higher than that calculated from parasitological data. Parasitological assessment (itself subject to various inaccuracies) nevertheless represents the proportion of successful inoculations, therefore the value of very accurate man-biting indices is questioned. Perhaps some type of human bait trap will give satisfactory information.

However, an overestimation of the man-biting index will always be present because usually only adult baits can be used whether in direct capture or in net traps, while there is some indication that infants and young children receive fewer bites than adults.

The average number of blood fed mosquitos found in spray catches may be divided by the number of occupants to give an index of number of bites per occupant, but is not considered to be a good substitute because:

- Some fed females may leave the house during the night.
- Females which have fed outside may enter the house.
- Houses containing animals could give misleading results unless correction is made by precipitin testing.
- Mosquitos concentrate in unsprayed structures in sprayed areas.
- It is often difficult to establish how many occupants are in the house.

Unless precise observations indicate that the biting activity of the vector is concentrated in part of the night, partial night catches should not be encouraged.

Sampling of house-frequenting and outdoor-resting populations

House resting observations in unsprayed houses involving ovarian stage classification will give indications of exophily. Such observations would be confirmed by the use of window traps. Final confirmation comes from studies of the outside resting population. While most species of Anopheles show some degree of endophily, in such species as A. balabacensis this is limited to a period when houses are visited to obtain a bloodmeal and little more.

The outside-resting sites of most Anopheles fall into two categories:

- in vegetation, e.g., A. aquasalis, A. pharoensis,
- in "solid background shelters", e.g., A. gambiae, A. funestus, A. albimanus.

Direct searching in the first type of resting site is difficult and usually unproductive. Drop-nets may be used, but unless the vegetation is exceptionally uniform, it is impossible to use them for quantitative sampling.

In the second type, direct searching can be quite productive and useful qualitative data can be obtained. Material for precipitin testing can be obtained likewise. Artificial shelters, such as box-shelters or pits, are frequently much more productive and can provide quantitative data, but they operate in competition with natural resting sites (and with houses if placed within villages) the availability of which varies with the seasons. Consequently, the population indices they provide may not correspond to changes in natural population levels.

Pit-shelters, which must be dug in shaded situations, are often the most productive, but cannot be used in low-lying sites. Of the other types of shelter, the choice of trap will depend on the materials and facilities locally available. Hard and fast rules cannot be laid down for the number of traps needed. But something of the order of 20 in any particular area should be aimed at, bearing in mind that only a few of the traps are likely to yield workable numbers of mosquitos.

As regards the siting of artificial shelters, there are three basic situations:

- near hosts, i.e., in villages or near animal sheds,
- in open country round villages,
- near breeding sites.

All of them are likely to yield different samples in terms of gonotrophic stages, which makes it almost impossible to estimate the overall composition of the outdoor resting population. Their use in assessment, however, is important both in providing an additional measure of the impact of control schemes on mosquito populations and in obtaining material for precipitin tests. Collections of outside resting species will be necessary to establish the human blood index.

The value of routine sampling of outdoor resting populations is debatable. The consensus of opinion is that this sampling should not be carried out on a routine basis but should be made only periodically.

A. balabacensis is difficult to find in outside resting places. With this species, it is considered sufficient to estimate the man-biting index and parous rate and to determine trap mortality in the evaluation of the effect of attack measures. It should be ensured, as in other cases where outlet traps are used, that the number of trap premises is increased after spraying.

Sampling of blood-meal smears for precipitin testing

The problem of sampling bloodmeals is a difficult one. Representative sampling is the key to accurate assessment of host preference. Samples may be taken either in the vicinity of potential hosts or on a "randomized" basis, which means in effect anywhere they can be found outdoors.

The first is valuable for estimating the extent of deviation to domestic animals under village conditions. To obtain valid data, both indoor and outdoor samples are required. Because one does not know the proportion of the total fed population resting indoors and outdoors respectively, it is not possible to estimate the distribution of feeds in the whole population. Nevertheless, material for precipitin tests collected in this way does give a valuable picture of feeding in the domestic environment, which often represents the main source of food for mosquitos from a considerable area around.

Material from non-domestic situations is usually scanty and mainly of interest in demonstrating the spectrum of hosts acceptable to a particular Anopheles. While baseline data on this characteristic are an essential part of any malaria programme, the collection of such data is not considered necessary on a regular basis. Information collected in different biotopes should be recorded separately and not pooled.

In relation to the problem of sampling bloodmeal smears, two suggestions are made:

(i) Improvement of the service provided by WHO, by better planning and inclusion of ecological information in and around the sites of collection of bloodmeal smears as has recently been introduced.

(ii) The need for research is emphasized. Two approaches could be used.

(a) The selection of representative man/animal combinations. One suggestion is to make observations in 5 houses, 3 of which are selected as average and 2 of which are opposite extremes. As complete a catch as possible should be aimed at and collection of a sufficient sample from the outdoor biotopes should be attempted when dealing with a vector showing an exophilic tendency.

(b) A direct comparison of all types of baits available (i.e., range of hosts) by means of a selected trapping method (Wharton, 1951). Care would be needed in the positioning and spacing of such baits in relation to the vector source. Switching of host position is an essential part of the technique.

Sampling for age-grouping and determination of duration of the gonotrophic cycle

Age-grading should be confined to those mosquitos coming to bite man because of the difficulty in obtaining samples representative of the age composition of the whole population. In this connexion, the use of light-traps inside houses where occupants are protected from mosquito bites may offer a good source of material from this population. Alternatively, collections with bed-nets with partitions to prevent feeding on the enclosed bait could also provide a suitable sample for parous rate determination.

In considering the value of parous rate determinations as distinct from direct age composition determinations (e.g., by counting actual members of follicular dilatations), the following factors should be borne in mind:

(a) Doubts about the precise value of k (the interval between bloodmeals) in the expression: proportion parous = p^k.

(b) Short-term fluctuations in age composition due to the varying rate of production of young female anophelines. This means that parous rates have to be averaged out over weeks or months, during which time the insecticidal effect may have changed considerably.

(c) The presence in some species of pre-gravid females feeding twice in the first cycle, which inflates the proportion of nullipars in the whole population.

(d) The fact that, where outdoor animal-biting and resting occurs on any scale, the parous population is recruited from two groups: those that fed on man previously and had, therefore, been exposed to insecticide, and those that fed on cattle previously and had not been at risk. With effective house treatment, there should be few survivors from the former group but many from the latter. A high parous rate may, therefore, simply be the result of a high degree of animal feeding outdoors even though a satisfactory kill of the man-biting and transmitting fraction of the population may have been achieved.

More field research on the interval between bloodmeals is required. The use of release and recapture techniques for the direct determination of this parameter is essential.

Prospects for mechanical devices

Consideration may be given to the use of light-traps (white and ultraviolet), particularly when used inside houses where the occupants are protected from mosquito bites. Garrett-Jones & Magayuka (unpublished information, 1973), who used this technique combined with window traps, found that such protection increases the size of the catch considerably. This method should be compared in efficiency with both all-night man-biting catches and the use of human-bait net traps.

Mention may be made of the tendency of the light-traps to attract other insects, sometimes in very large numbers, and subsequent difficulties in separating the target insect. A recent development is the use of updraft rather than downdraft types of traps which tended to reduce the number of heavier types of insects (Wilton & Fay, 1972).

The operational difficulties revolve around failures in power supply, maintenance and battery weight.

Sampling larval populations

Difficulties inherent in larval sampling are notorious because of the diversities of the breeding places and non-random distribution of larvae in breeding places.

Where sampling is for life-table construction, differences in sensitivity of larval instars to disturbance produce biased samples. Life-table construction is an important element in planning integrated control measures.

BIBLIOGRAPHY

Bishop, O. N. (1968) Statistics for biology, Edinburgh, Longmans

Christie, M. (1956) The statistical treatment of the sporozoite rate in anopheline mosquitos, Ann. trop. Med. Parasit., 50, 350

Grab, B. (1968) Practical applications of standards for response of malaria to early attack measures and considerations on the requirements to ensure statistical confidence in them, WHO/MAL/68.653 (Unpublished mimeographed document)

Logan, A. J. (1953) The Sardinian Project; An experiment in the eradication of an indigenous malaria vector. The Johns Hopkins Press, Baltimore

Knight, K. L. (1964) Quantitative methods for mosquito larval surveys, J. med. Ent., 1, No. 1, 109-115

Mouchet, J. (1972) Méthodes et indices employés pour les enquêtes nationales. Evaluation des ces enquêtes du point de vue écologique, WHO/VBC/72.333

Odetoyinbo, J. A. (1969) Preliminary investigation on the use of a light-trap for sampling malaria vectors in The Gambia, Bull. Wld Hlth Org., 40, 547

Service, M. W. (1971) Studies on sampling larval populations of the A. gambiae complex, Bull. Wld Hlth Org., 45

Shalabi, A. M. (1971) Sampling of outdoor resting populations of A. culicifacies and A. fluviatilis in Gujarat State, India, Mosq. News, 31, 68

Smith, J. M. (1968) Mathematical ideas in biology, Cambridge, University Press

Swaroop, S. (1966) Statistical methods in malaria eradication, Wld Hlth Org. Monogr. Ser., 51

Rayevsky, G. E. (1942) Methods of epidemiological assessment of entomological observations, Med. Parazit. (Mosk.), 11, 46

Wharton, R. H. (1951) The habits of adult mosquitos in Malaya. I. Observations on Anophelines in window-trap huts and at cattle-sheds. II. Observations on Culicines in window-trap huts and at cattle-sheds, Ann. trop. Med. Parasit., 45, 141-160

Wilton, D. P. & Fay, R. W. (1972) Air flow direction and velocity in light-trap design, Ent. Exp. & Appl., 15, 377-386

CHAPTER 4. RECORDING AND REPORTING OF DATA

CONTENTS

	Page
RECORDING OF ENTOMOLOGICAL DATA	110
REPORTING OF ENTOMOLOGICAL DATA	110

CHAPTER 4

RECORDING AND REPORTING OF DATA

Recording of entomological data

Correct recording is essential for the interpretation of entomological data. Special care must be taken in recording the numbers of insects collected and processed, and in recording data referring to the methods used and the environmental conditions in which the collections were made, depending on the objectives of the investigations (details can be found in the section dealing with entomological methods).

Recording of data includes the following steps:

(a) Recording of data in the field; for such purposes special forms or field books are used for recording the data as soon as it is collected.

(b) Recording of data in the laboratory after the processing of the material; the recording of processing could be inserted on the same forms on which the field data have been recorded. A number of tentative forms for recording are included in this manual (Forms 1-5), and these may be adapted in order to suit local needs.

Apart from the recording of data on the forms, it is particularly useful to record the data on graphs, either as crude data or after processing. The graphical presentation, when correctly done, has the advantage of illustrating at once the trends of different phenomena or situations in time and space, taking into account the influence of various factors. It should be noted that graphs should not include too many lines as this makes immediate interpretation difficult.

Reporting of entomological data

Reports of entomological data should contain the following features:

- A description of the objectives and conditions of work, also the methods and techniques applied.

- Results obtained by different types of investigation at representative sites.

- Comments and interpretation of data.

- Conclusions and recommendations (if required).

The comments and discussion on the value and significance of various entomological data can be supported by figures, and the interpretation may be based on the data collected by several methods together with environmental influences occurring before and at the time of the entomological data collection.

Monthly reports consolidate the data collected during the month and are of value for adjusting the activities during the following month if necessary. Quarterly reports offer a greater possibility for interpretation of data, and the trends in entomological aspects appear more clearly than in the monthly report, so permitting partial conclusions to be drawn. Annual reports are also very important as all seasonal variations have been encountered, enabling a more realistic perspective and trends of the phenomena to be studied in relation to the evolution of the disease and of the value of measures applied. This is very important for the evaluation of the situation as well as for adapting, and if necessary improving, the planning for entomological investigations and operations during the next year.

FORM 1. LARVAL COLLECTIONS

Form No. ....

Country ....
Region ....
Area ....
Zone ....
Section ....
Locality ....
Altitude[2] ....
Location on the map (No. ....)

Unsprayed/sprayed area ....
Date of last residual spraying ....
Insecticide ....
Date of last larviciding ....
Larvicide .... Quantity/ha

Collection record:

Name of collector ....
Date of collection ....
Time of collection ....

Date (period)	Type of breeding place and approx. surface[1]	Distance from nearest house[2]	Water characteristics			Light[4]	Vege-tation[5]	Method of collec-tion[6]	Breeding places investig.		No. of dips	No. of larvae[7]	Larvae per dip[7]	Species identified and No. of larvae[7]							
			Quiet running	Clear turbid	Depth approx.[3]				No.	Positive				Total	per dip	Total	per dip	Total	per dip	Total	per dip

[1] Spring, stream, river bed, irrigation canal, swamp-like seepage, hoofprint, rain pool, spring pool, spring well, artificial containers (others to specify).
[2] If significant.
[3] Shallow (less than 1 metre deep), deep (more than 1 metre deep).
[4] Sunlight, semi-shade, shade, deep shade.
[5] Emergent, floating, prevalent species if identified.
[6] Dipping, larval net, (others to be specified) and approximate area of investigated surface.
[7] In brackets, larvae of III and IV stage when detailed investigations are made.

This form could be used in preliminary surveys, collection of baseline data and evaluation of anti-vector operations, for recording the results of investigations of each breeding place, or for the results obtained from several breeding places belonging to the same type.

FORM 2. COLLECTION AND EXAMINATION OF RESTING ADULT MOSQUITOS
(for baseline and follow-up studies)

Country ....
Region ....
District (area) ....
Locality and No. on the area map ....
Date of last spray and insecticide ....

Form No. ....
Team No. ....

Method of Coll.	Date of Coll.	Habitat investigated					Vectors collected																	Remarks[d]
		Type[a]	No.	No. pos.	No. occup. & spec.	DFBP[b]	A										A										
							♂	♀	Tot.	U	F	HG	SG	G	P/TD[c]	SP/TD[c]	♂	♀	Tot.	U	F	HG	SG	G	P/TD[c]	SP/TD[c]	

[a] Type of the habitat investigated, House = H; stable = S; outdoor vegetation = OV; holes = H, specify others when the case.
[b] DFBP - distance from breeding places if known.
[c] If such investigations are carried out.
[d] Mention under remarks: meteorological data, number of blood smears for precipitin tests (when collected) and other aspects which might influence the resting of mosquitos.
   P - parous, TD - total dissected, SP - sporozoites.

FORM 3. ADULT COLLECTION AND RESULTS OF PROCESSING COLLECTED MATERIAL
BY LOCALITY AND DATE OR PERIOD OF COLLECTION

Country ....
Region ....
Area ....
Altitude* ....

Form No. ....

Pyrethrum spray/or hand collection

Date	Locality	Altitude*	Habitat			A									Remarks
			Type	No. occup.	Ex/p	♂	♀	Tot.	U	F	HG-SG	G	P/TD	SP/TD	

Night bait collection

Date	Locality	Wind, rain, moon phase	Type of bait**	No. baits		Tot. hours		Total vectors		A									A								
				in	out	in	out	in	out	Total		BR		P/TD		SP/TD			Total		BR		P/TD		SP/TD		
										in	out	in	out	in	out	in	out		in	out	in	out	in	out	in	out	

* When necessary, and the village number on the map of the area.
** Mention also the hours of collection when less than whole-night collection is done (e.g.,19-21 hours etc.)

Moderate wind (MW), strong wind (SW), Light rain (LR), Heavy rain (HR), Cloudy (C), moon phase 1/4, 2/4, 3/4, 4/4.
Type of baits - Human (H), animal; B = buffalo, D = donkey, H = horse, C = cattle, P = pig
Dissection for parity or sporozoite rates only if necessary.
BR = biting rate per man/hour or night.

FORM 4. TRAP COLLECTION

Form No. ....

Country ....
Region ....
Area ....
Village(s) No. ....
Sprayed/unsprayed ....
Date of last spray* ....

Capture station No. ....
Sprayed/unsprayed ....
Insecticide and conc./m^2 .....
Date of last spray* ....
Date of observations ....

1	2	3	4			5				6	7	8					
Trap collection	Vectors	Total	Vectors collected			Abdominal appearance				P/TD*	SP/TD*	24-hours' survival rate (females)					
			Alive	Dead	Total	UF	F	HG	SG	G			Dead		Alive		%
													UF; F	HG-G	UF	HG-G	
Type of Trap	A																
	A																
	A																
	A Total																
	Vectors per trap																
Day-time spray (hand)* catch in huts with traps	A																
	A																
	A																
	A																
	Total																
	Total per room																
Grand total per room																	

9
No. of other species collected in trap ....
....

10
No. of other species collected by spray (hand) ....
....

Remarks
Presence of fire during night
.... Yes No ....
Moonphase 1/4, 2/4, 3/4, 4/4
Wind** ...... Rain** ......

* If the case.
** Heavy or moderate.

If this table is used for recording the data from bait traps, specify the baits.

This form can be used for recording collections in local houses, in experimental trap huts or other trapping systems by filling only the appropriate columns.

FORM 5. DENSITY OF BITING ANOPHELES

Form No. .....

Country ....
Region ....
Zone ....
Area ....
Locality ....
Capture station No. ....
Distance from nearest breeding place ....

Vector species 1.A ....
Vector species 2.A ....
(Additional columns might
be added if more vectors
are found)

Date of last spray ....
Moon phase 1/4, 2/4, 3/4, 4/4 ....
Date (period) of investigation ....
Type of room where indoor collection
is made ....
Spraying status of the indoor ....

Time of day	Inside						Outside*						Parous (Total dissected)				Sporozoites (Total dissected)				Remarks** (F)
	No. baits (F)	Total coll. anoph. (F)	No. of vectors		Vectors per bait night		No. baits (F)	Total Coll. (F)	Vectors per bait night		No. of vectors		Inside		Outside		Inside		Outside		
													Vector		Vector		Vector		Vector		
			1	2	1	2			1	2	1	2	1	2	1	2	1	2	1	2	
18-19																					
19-20																					
20-21																					
21-22																					
22-23																					
23-24																					
24-01																					
01-02																					
02-03																					
03-04																					
04-05																					
05-06																					
Total																					

* Distance from the indoor capture station.
** Mention: gravid females coming to feed; newly emerged females (if identified) etc. and T°C, RH% when recorded.
( ) Total dissected.
W = Wind; MW = Moderate wind; SW = Strong wind; R = Rain; MR = Moderate rain; HR = Heavy rain.
People outside in the area, outdoors during the capture time. Domestic animals in a radius of 25 m.
F - columns marked (F) should be completed in the field; the rest should be completed in the laboratory.
Dissection for parity and infectivity only when necessary.

Country ....
Region ....
Area/district ....
Locality ....

FORM 6. BIO-ASSAY TESTS

Form No. ....

Spraying status, insecticide ....
Quantity per square metre ....
Name of investigator ....

Contact bio-assay ☐
Bio-assay for fumigant effect ☐

Date of test	Name of Householder or Code No.*	Type of place	Date last spray	Quality of spray	Mosquito density**	Time of exposure	Site of exposure	No. cages or cones	No. mosquitos***	Knock down and/or dead mosquitos		Controls		Corrected mortality	T°C during		RH% during	
										At end of exposure	24 hours	No.	Dead 24 h		Exp.	24 h	Exp.	24 h

* For monthly or quarterly reports mention the locality.
** Mention the mosquito species found resting inside in the morning.
*** Mention species and feeding status; UF - unfed; BF - blood fed; SF - sugar fed; lab.colony/wild; (L);/(W).

The tabulation or graphic representation of entomological findings for a whole year should be presented month by month, and the interpretation of their significance should be made while taking into account: (a) the monthly variation of the environmental conditions, e.g., temperature, humidity, rains and any other related phenomena which normally have an influence on mosquitos and on the variability of entomological data, and (b) the monthly variations of the disease in the localities and areas where the entomological investigations have been carried out.

Both types of information should be plotted on semilogarithmic paper in order to illustrate the trends and facilitate correlation. In the interpretation of entomological data in a programme where antivector measures are applied, consideration has to be given both to the impact of the antivector measures on the vector population and to the consequences of the antivector measures on transmission. In the first case the entomologist has to assess the performance of the insecticide in controlling the vector population and its contact with man and, in the second case, to contribute to the global epidemiological evaluation of the antivector measures, by estimating the variations of entomological inoculation rate.

Percentages and ratios should be calculated only on statistically significant numbers.[1]

---

[1] Swaroop, S. (1966) Statistical Methods in Malaria Eradication, Wld Hlth Org. Monogr. Ser., 51.

CHAPTER 5.   INTERPRETATION OF ENTOMOLOGICAL AND RELATED DATA

CONTENTS

	Page
1. METHODS OF INTERPRETATION OF DATA	120
2. EXAMPLES OF JOINT PARASITOLOGICAL/ENTOMOLOGICAL INTERPRETATION	123
ANNEX 1. OUTLINE FOR ANALYSIS, INTERPRETATION AND PRESENTATION OF ENTOMOLOGICAL AND RELATED DATA	149
BIBLIOGRAPHY	157

CHAPTER 5

INTERPRETATION OF ENTOMOLOGICAL AND RELATED DATA

1. METHODS OF INTERPRETATION OF DATA

1.1 SIMPLE COMBINED EXPRESSION OF ENTOMOLOGICAL AND PARASITOLOGICAL INDICES

While some contributions by the entomological services are purely descriptive, requiring little in the way of interpretation (e.g., geographical distribution and seasonal variations in vector density), others, such as those related to the actual or potential force of infection and the results of susceptibility testing present greater problems when an attempt is made to assess their significance. It is essential that all entomological indices derived for epidemiological purposes be constantly related to the particular parasite reservoir involved. While for demonstration of long-term trends in a particular index the mean of several readings is usually sufficient, it should be remembered that, owing to the focal nature of malaria transmission, a reading at the extreme of a range can often be highly significant epidemiologically. Thus any adverse readings obtained for a particular index must always be considered on their own merits and not masked by combination with other readings to form a misleading average figure. A really sound appraisal of control measures requires the assessment of both parasitological and entomological trends. The parasitological trend is shown in changes of the parasite rate in the general population. The entomological trend can be expressed in simple terms without necessarily being expressed in a mathematical way. Graphical presentation of entomological and parasitological indices will facilitate this approach to interpretation, (Fig. 25 and 26).

Relative entomological indicators in this connexion are:

(i) A combination of man-biting index and sporozoite rate and/or parous rate, depending on vector species.

(ii) Numbers of mosquitos occurring in sprayed houses.

(iii) Changes in abdominal and ovarian stages in females captured in indoor spot-check catches and window trap catches.

(iv) Changes in mortalities in window traps and changes in susceptibility. A combination of these will give increased confidence and will also include an early warning element. Parallel parasitological and entomological data should be gathered from the same indicator areas, and in this connexion, parasitological data should be gathered more frequently than at annual intervals. An outline for the analysis, interpretation and presentation of entomological and related data is given in Annex 1.

In larvicidal programmes, entomological confirmation of trends in control will be given by daytime house-resting adult density estimations (Chapter 2, section 5). These data, prepared by both entomologists and parasitologists, should be graphically presented in their simplest form.

1.2 THE USE OF MATHEMATICAL EXPRESSIONS

The two main entomological contributions to a description of malaria transmission in a given area are the estimates of entomological inoculation rate and vectorial capacity.

1.2.1 Entomological inoculation rate

This can be summarized in the expression $\underline{h} = \underline{m} \times \underline{a} \times \underline{s}$ (where $\underline{h}$ is the number of positive bites received by one person in one night, $\underline{m}$ is the anopheline density relative to man, $\underline{a}$ is the man-biting habit of the anopheline species and $\underline{s}$ is the sporozoite rate in the biting population). Where two or more vectors are involved, their separate indices are summed to give an overall inoculation rate.

The aim of estimating the entomological inoculation rate is merely to assist in describing the actual source of infection existing in a given locality. It is the epidemiological complement of the parasitological inoculation rate derived from malaria incidence surveys of the human population. Though relatively easily determined in areas of high endemicity before control measures are applied, the entomological inoculation rate becomes difficult to quantify afterwards, when adult densities decline and sporozoite infections become rare.

Material for estimation of the entomological inoculation rate is sampled by night-biting catch or pyrethrum spray catch plus exit trap collection. It is felt that, compared to the actual situation, the latter provides a relatively more accurate approximation (although probably an under-estimate) and the former an over-estimate of inoculation rate owing to an expected exaggeration of man-biting density over the natural situation (see also section headed: Sampling for estimation of man-vector contact). A further divergence in comparability of data is due to the fact that the night-biting catch is made using adults as bait while the parasitological inoculation rate is usually based on surveys of infants. In highly endemic areas a further important source of variation in the sporozoite rate may arise from changes in the growth of, or the decline of, mosquito populations. Sometimes many vectors show marked instability in number depending on the availability of favourable breeding sites, and predominantly young or old populations occur simply through the operation of this factor. This can give rise to striking short-term variations in the sporozoite rate, and such changes may add to the difficulty of accurately assessing the effect of control measures. In determining infection rates, it is essential, therefore, to carry out observations at repeated intervals and at different seasons so as to allow for these natural changes in density, contact with infected man, and longevity (age composition). The need for adequate sampling and processing of material is essential.

Since confidence limits in determination of the sporozoite rate depend on the size of the sample dissected, the following table (Table 8) has been included to give some indication of the accuracy expected from different sample sizes. (For further discussion of this, see Christie, 1956).

It is normally found that there is a fairly large discrepancy between the entomological and parasitological inoculation rates, the former being consistently larger than the latter - perhaps of the order of sixty times. This difference, which may be due to various factors such as low viability or low numbers of sporozoites inoculated, the relatively immune status of the human hosts, etc., is that designated by Macdonald as the factor $\underline{b}$ which he adds to the above expression to describe the effective entomological inoculation rate: $\underline{h} = \underline{m} \times \underline{a} \times \underline{b} \times \underline{s}$.

It will therefore be recognized that the estimation of the entomological inoculation rate has little practical value per se, since the equivalent information can be more easily and accurately determined by serial parasitological surveys of the human population involved, the latter being the end result of the whole transmission process from the point of view of malaria control or eradication. Estimation of the entomological inoculation rate is therefore to be recommended in operational programmes only as a confirmatory complement to parasite surveys, or where it is not possible to carry out the latter.

TABLE 8. DISSECTION FOR SPOROZOITE RATE:
CONFIDENCE INTERVALS FOR OBSERVED PERCENTAGES AT 95% PROBABILITY LEVEL

Sample size	1.0	2.0	3.0	4.0	5.0	6.0	7.0	8.0	9.0
50	–	0.005-10.0	–	0.045-14.0	–	1.2 -16.0	–	2.2 -19.0	–
100	0.02-5.45	0.24 - 7.04	0.62-8.52	1.10 - 9.93	1.64-11.29	2.23-12.61	2.86-13.90	3.51-15.16	4.19-16.40
200	0.12-3.57	0.54 - 5.05	1.10-6.42	1.74 - 7.73	2.42- 9.01	3.13-10.25	3.88-11.47	4.64-12.67	5.42-13.85
300	0.20-2.90	0.73 - 4.31	1.38-5.62	2.08 - 6.89	2.82- 8.12	3.59- 9.32	4.38-10.51	5.19-11.67	6.01-12.83
400	0.27-2.54	0.86 - 3.91	1.56-5.19	2.30 - 6.42	3.08- 7.62	3.88- 8.80	4.70- 9.96	5.53-11.11	6.38-12.25
500	0.32-2.32	0.96 - 3.65	1.68-4.90	2.46 - 6.11	3.26- 7.30	4.08- 8.46	4.92- 9.61	5.77-10.74	6.64-11.86
600	0.36-2.17	1.03 - 3.47	1.78-4.70	2.58 - 5.90	3.39- 7.07	4.23- 8.21	5.09- 9.35	5.95-10.47	6.83-11.59
700	0.40-2.05	1.09 - 3.34	1.86-4.55	2.67 - 5.73	3.50- 6.89	4.35- 8.03	5.22- 9.15	6.09-10.27	6.98-11.38
800	0.43-1.97	1.14 - 3.23	1.93-4.43	2.75 - 5.60	3.59- 6.75	4.45- 7.88	5.33- 9.00	6.21-10.11	7.10-11.21
900	0.45-1.89	1.19 - 3.15	1.98-4.34	2.81 - 5.50	3.67- 6.64	4.53- 7.76	5.42- 8.88	6.31- 9.98	7.21-11.07
1 000	0.48-1.84	1.22 - 3.08	2.03-4.26	2.87 - 5.41	3.73- 6.54	4.60- 7.66	5.49- 8.77	6.39- 9.86	7.29-10.95
1 500	0.56-1.65	1.35 - 2.85	2.19-4.00	3.06 - 5.13	3.95- 6.23	4.85- 7.33	5.76- 8.42	6.67- 9.50	7.60-10.57
2 000	0.61-1.54	1.43 - 2.72	2.29-3.85	3.18 - 4.96	4.08- 6.05	5.00- 7.14	5.92- 8.21	6.84- 9.28	7.78-10.35
2 500	0.64-1.48	1.48 - 2.63	2.36-3.75	3.26 - 4.85	4.17- 5.93	5.10- 7.01	6.03- 8.08	6.96- 9.14	7.90-10.20
3 000	0.67-1.43	1.53 - 2.57	2.41-3.68	3.32 - 4.77	4.24- 5.85	5.17- 6.92	6.11- 7.98	7.05- 9.03	8.00-10.09
3 500	0.69-1.39	1.56 - 2.53	2.46-3.62	3.37 - 4.71	4.30- 5.78	5.23- 6.84	6.17- 7.90	7.12- 8.95	8.07-10.00
4 000	0.71-1.36	1.58 - 2.49	2.49-3.58	3.41 - 4.66	4.34- 5.73	5.28- 6.79	6.22- 7.84	7.17- 8.89	8.13- 9.93
4 500	0.73-1.34	1.61 - 2.46	2.52-3.55	3.44 - 4.62	4.38- 5.68	5.32- 6.74	6.27- 7.79	7.22- 8.84	8.18- 9.88

Observed percentage

1.2.2   Vectorial capacity

Vectorial capacity, which is the same thing as the basic reproduction rate of malaria expressed on a daily basis, is another concept which may be used to describe the force of infection in a particular epidemiological situation.

Since measurement of the sporozoite rate is not involved, the vectorial capacity may be estimated whether or not transmission is actually taking place, and comparison of vectorial capacities before and after the implementation of anti-vectorial measures may be taken as a measure of the "impact" of the latter on the vector population and thus on transmission.

Thus perhaps the main use of the vectorial capacity concept is in an attempt to monitor transmission potential in areas where no transmission is taking place but which are both vulnerable <u>and</u> receptive.

Since the expression for vectorial capacity

$$\frac{ma^2 p^n}{-\log_e p}$$

is composed of four variables, three of which may change at short notice and the fourth of which changes with temperature, it is apparent that frequent re-determination of its constituents would be necessary if it is used as a reference for monitoring purposes over a long term. Even this assumes that each variable may be accurately determined - which is not the case so far.

"Difficulties of representative sampling and of interpretation continue to pose serious obstacles to the operational assessment of each one of the factors of vectorial capacity" (Garrett-Jones & Shidrawi, 1969).

The value of p, the probability of daily survival, is particularly difficult to assess, although some progress is being made by the use of developing malaria and filaria infections in age-grading. This, however, still does not facilitate the estimation of p in areas of lower disease and endemicity. At present, p is usually estimated from determinations of parous rates which need to be made on adequate numbers and over sufficient time to minimize the effect of short-term fluctuations in population composition (see Fig. 19, 20 and 21). These requirements militate against the successful application of this method as a short-term analytical tool. Other age-grading techniques such as that of Schlein & Gratz (1974), based on changes in skeletal apodeme structure, or that of Spencer, based on visual estimation of the amount of ovarian pigment present, remain to be fully evaluated. Once control measures are applied, numbers become a limiting factor in the use of vectorial capacity as an index.

2.   EXAMPLES OF JOINT PARASITOLOGICAL/ENTOMOLOGICAL INTERPRETATION

2.1   AN EXAMPLE OF INTERPRETATION OF BASELINE DATA

As an example of joint parasitological/entomological interpretation of epidemiological data, an extract is given below of a recent evaluation of data provided by a field research project in Kenya. The data referred to are baseline data acquired during Phase I, or the pre-spray period, of an insecticide trial.

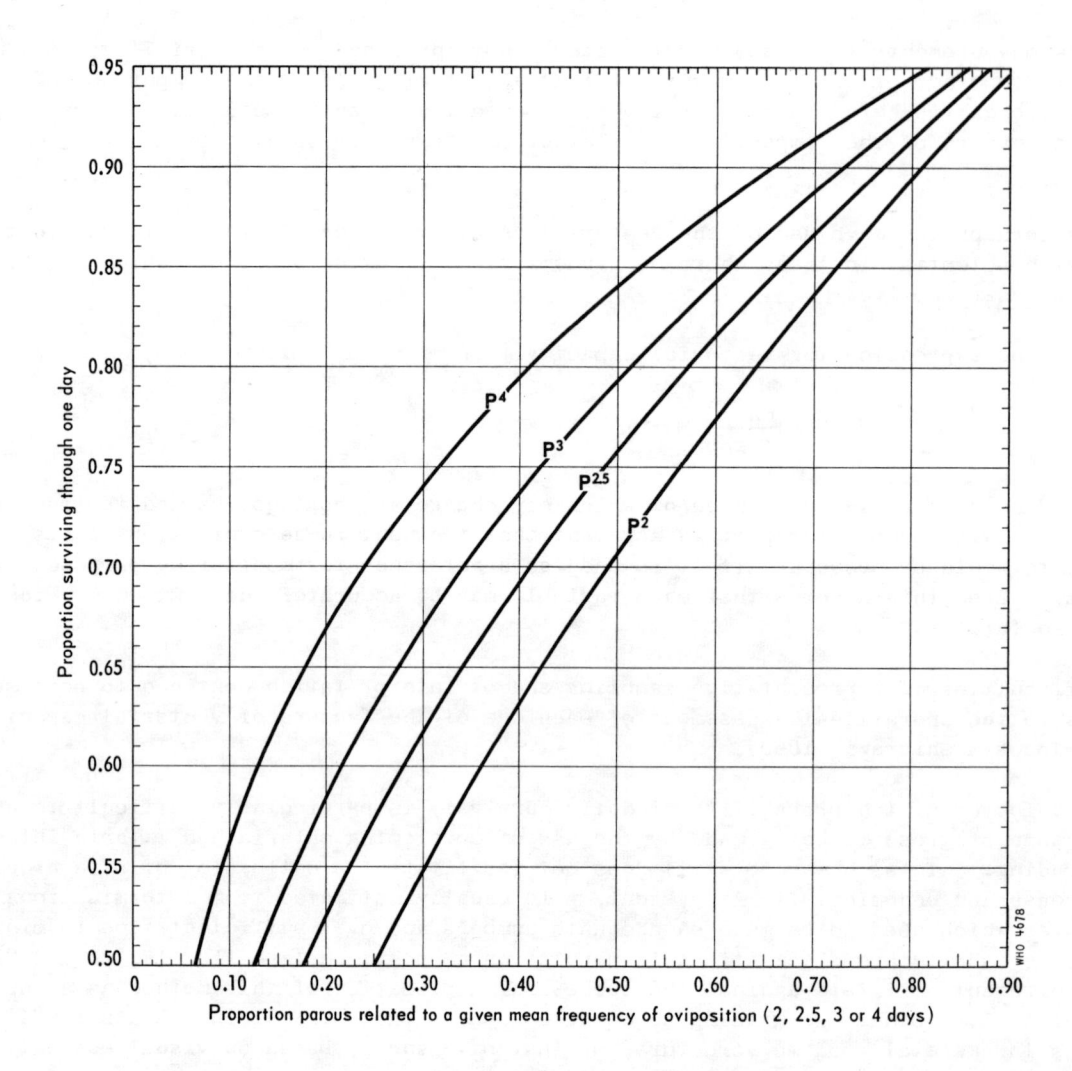

Fig. 19. Curves for deriving the proportion surviving through one day (p) from observed proportions parous representing p^2, $p^{2.5}$, p^3 or p^4.

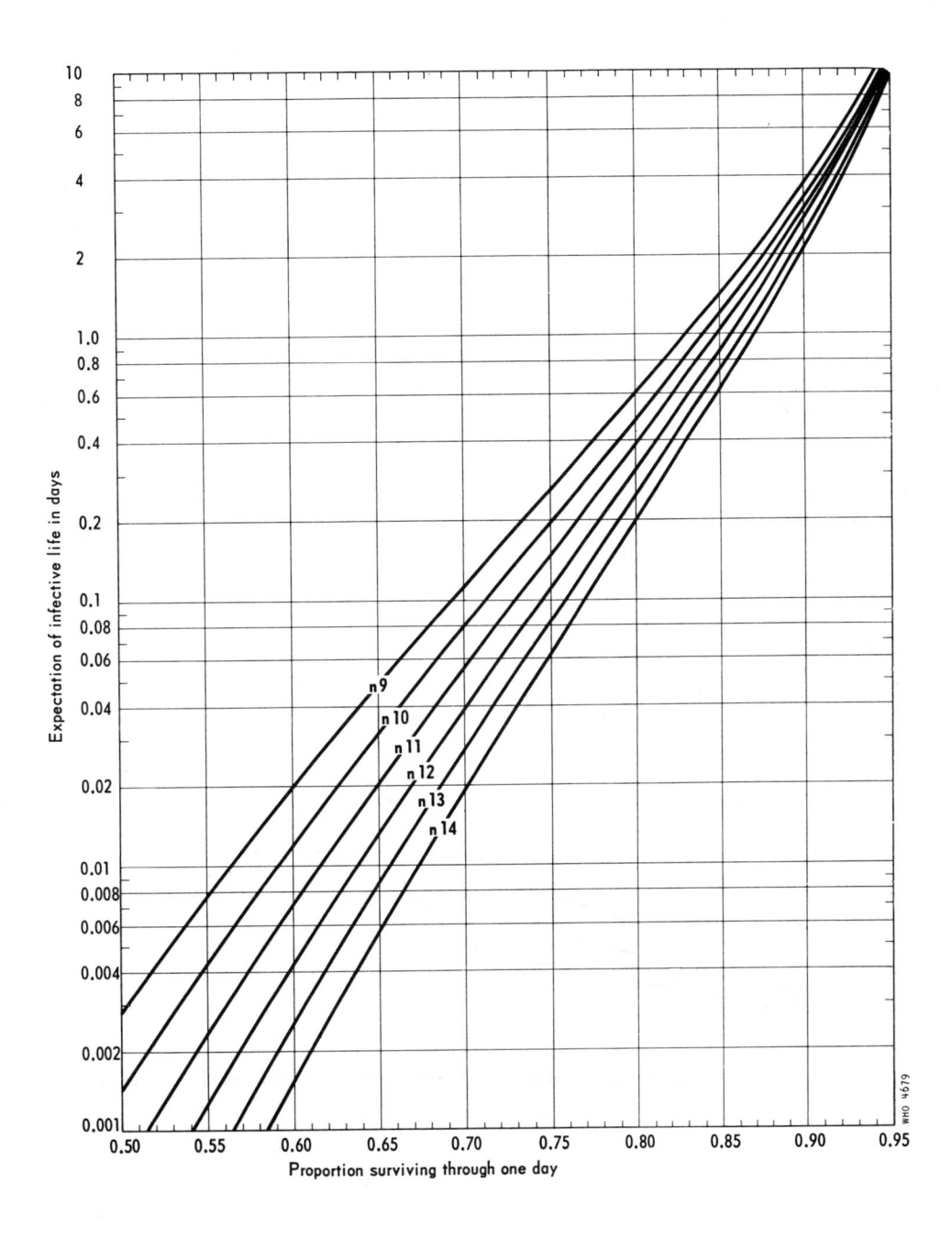

Fig. 20. Curves for deriving the expectation of infective life of the vector from known values of p, for sporogonic periods of 9 to 14 days in the parasite: the proportion surviving through one day is derived from Fig. 19.

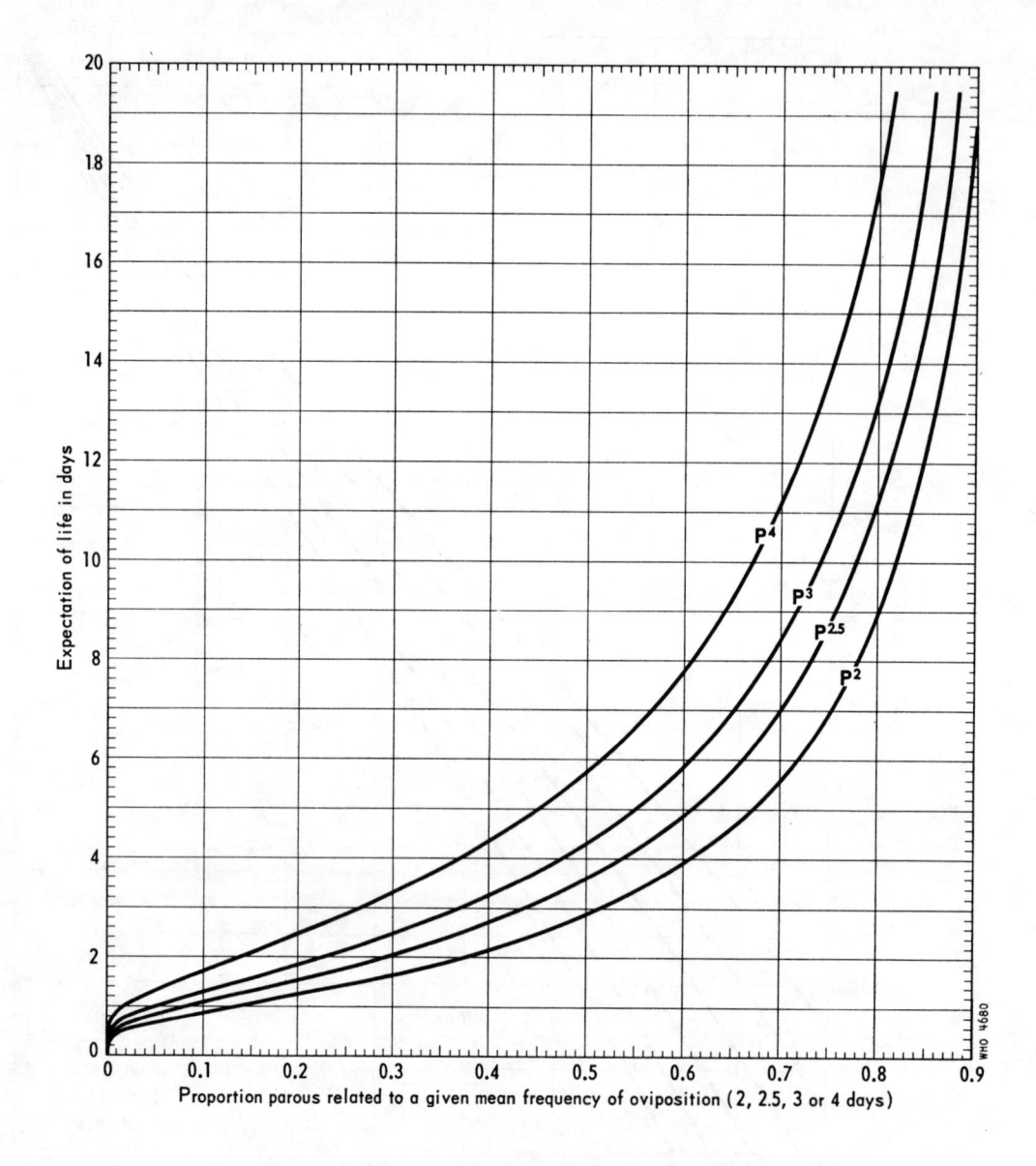

Fig. 21. Curves for deriving the expectation of life from observed proportions parous representing p^2, $p^{2.5}$, p^3, or p^4.

During Phase I of this project, efforts were made, under strict supervision, to obtain representative samples of the greatest possible size.

Prevalence variations of the disease, as well as parasite species distribution and gametocyte rates, have been estimated on a total of 8506 blood examinations, whereas incidence trends were established by the monthly follow-up of an average of 630 infants (maximum 666, minimum 568) from July 1972 till July 1973.

Adult vector densities indoors were established by a monthly average of 452 entomological observations throughout the study areas. In all, 254 521 malaria vectors were collected, identified by species and classified according to abdominal appearance during the period January 1972 - July 1973. Larval vector densities were established monthly in a minimum number of 10 sectors per month, during which 114 021 dips were made over the period February 1972 - July 1973.

A total of 94 918 gland dissections for detection of sporozoites were carried out averaging 5000 per month (maximum 8124, minimum 860) on which monthly rates were established. The direct estimation of anopheline daily survival rate was based on a total of 7614 A. gambiae and 10 723 A. funestus caught in exit traps. The man-vector contact was estimated on a total of 933 man-bait nights, the collections being carried out throughout the observation areas from February 1972 to July 1973 (indoors) and from June 1972 to July 1973 (outdoors).

The parity rates of both A. gambiae and A. funestus, collected indoors, were established on a total of 10 769 ovarian observations, with a monthly average of 331 A. gambiae and 266 A. funestus (from February 1972 to July 1973). Trophic preferences of A. gambiae and A. funestus were evaluated in 1972 on totals of 2903 and 2577 specimens, respectively, on which precipitin tests were performed, indicating a human blood index of 0.95 for A. gambiae and 0.99 for A. funestus.

Since both the human and anopheline samples were relatively large, it is possible to give a reasonably confident description of the malaria situation and of its variations during Phase I.

2.1.1 The malaria situation

A given malaria situation, under natural conditions or otherwise, may be described by reference to:

- the malaria infection load in a population and its variation according to age group;
- the daily inoculation rate derived from parasitological data;
- the daily inoculation rate derived from entomological data.

Evaluation of the infection load under natural conditions has been carried out by means of prevalence surveys in June/July 1972 and in December 1972 - January 1973.

This information, based on the longitudinal blood examination of clusters of population above one year of age (200 thick blood fields for each slide) shows that malaria can be considered as stable throughout the evaluation and comparison areas, although some seasonal variations are noted. The parasite rates were 64.7% in June/July 1972 compared with 58.1% in December 1972/January 1973 for the evaluation area and 64.5% against 53.2% respectively in the comparison area. The data collected show that the malaria parasite rate decreases with advancing age as expected in hyper/holoendemic conditions. For example, the results of the first prevalence survey carried out in June/July 1972 in the evaluation area indicated a parasite rate of 100% in the age group 1-2 years, whereas the parasite rate was only 39.4% in the age group above 45 years (as shown in the tabulation below). The method of placing parasite rates (PR) and parasite densities (PDI) parallel is a realistic way of estimating the immunity trends of a population with increase in age.

Area	Sur-vey	Age groups						
		1-2	2-4	5-9	10-14	15-24	25-44	45+
Evaluation PR	1	100.00	83.4	89.0	86.8	60.3	42.7	39.4
PDI	-	3.03	2.63	2.18	1.62	1.44	1.29	1.20
ML (malaria load)		3.03	2.19	1.95	1.40	0.86	0.55	0.44

It could also be noted that in multiplying parasite rates expressed as a proportion by parasite densities for each age group it is possible to estimate roughly the average malaria load (ML) of the various age groups.

At each survey the parasite formula demonstrated a large prevalence of P. falciparum (69%) followed by P. malariae around 25.2% and P. ovale (5.6%).

Of value in assessing the potential infectivity to mosquitos of a given population is the point prevalence of gametocyte carriers among representative samples of population. In the period July 1972 - May 1973, the average P. falciparum gametocyte rate was 19% with point prevalence variations between 9.4% and 32.2% according to surveys and age group considered. In other words, as an average, it might be considered that 1 out of 5 susceptible mosquitos biting any given person could potentially become infected with P. falciparum. In addition, there was a relatively high percentage of persons with P. malariae and P. ovale gametocytes.

The incidence of primo-infections in infants, based on the monthly follow-ups, is shown in Table 9. From the last line of Table 9 the average daily rate of incidence in susceptible infants may be calculated. This is the daily effective inoculation rate (h in the expression $h = mabs$) and equals 0.0083. The age distribution of disease acquisition in a cohort of 1000 newborn may be established (Table 10) and a similar age distribution showing "apparent" disease acquisition may also be derived from the infant parasite prevalence data (Table 11). From the latter data the resulting average daily inoculation rate is 0.00605. By comparing the inoculation rates obtained from the curves drawn from these data (Fig. 22), one may derive a value for apparent recovery rate operating at any particular age and modified by superinfection (Table 12, Fig. 23).

The entomological inoculation rate may be calculated on the basis of the man-biting or the pyrethrum spray plus exit trap collections. The monthly values for the combined A. gambiae and A. funestus inoculation rates derived in these two ways are given in Tables 10, 11, 12 and Fig. 24. Here, $h = mas$ since the question of the infectivity of a "positive" bite does not arise.

TABLE 9. THE DISTRIBUTION OF PRIMO-INFECTIONS ACCORDING TO AGE AND NUMBER EXAMINED IN BOTH EVALUATION AND COMPARISON AREAS

	Infant age in months											
	0	1	2	3	4	5	6	7	8	9	10	11
Number of susceptibles examined	507	443	320	222	131	79	39	21	16	6	4	1
Number of primo-infections	12	82	73	64	33	26	11	5	3	1	2	0
Monthly incidence rate (%)	2.4	18.5	22.8	28.8	25.2	32.9	28.2	23.8	18.8	16.7	50.0	0.0

From these data the following values may be calculated:
average monthly incidence, 24.8
average daily incidence, 0.0083

TABLE 10. THE AGE DISTRIBUTION OF DISEASE ACQUISITION IN A COHORT OF 1000 NEWBORN

Age in months	Number susceptible	Incidence rate %	New cases in cohort	
			Per month	Cumulative
0-	1 000	2.4	24	24
1-	976	18.5	181	205
2-	795	22.8	181	386
3-	614	28.8	177	563
4-	437	25.2	110	673
5-	327	32.9	108	781
6-	219	28.2	62	843
7-	157	23.8	37	880
8-	120	18.8	23	903
9-	97	16.7	16	919

TABLE 11. INFANT PREVALENCE SURVEYS - EVALUATION AREA
(FOUR SURVEYS COMBINED)

Age in months	Number examined	Number positive	% positive
0-	151	2	1.3
1-	168	23	13.7
2-	188	63	33.5
3-	168	75	44.6
4-	175	83	47.4
5-	204	126	61.8
6-	178	103	57.9
7-	179	110	61.5
8-	201	120	59.7
9-	159	99	62.3
10-	168	102	60.7
11-	170	115	67.6
Total	2 109	1 021	48.4

Visual estimate at end of month	
From prevalence	From age specific incidence
7.0	10.0
24.0	30.0
37.0	47.0
47.5	61.5
55.0	73.0
60.5	81.0
63.5	86.0

TABLE 12. VALUES OF DAILY INOCULATION RATE (h) BASED ON INFANT INCIDENCE DATA AND OF (d) BASED ON INFANT PREVALENCE DATA DERIVED BY APPLICATION OF THE FORMULAE $x = 1-e^{-ht}$ AND $x = 1-e^{-dt}$ RESPECTIVELY; TOGETHER WITH VALUES OF "r" THE APPARENT RECOVERY RATE (RECOVERY RATE MODIFIED BY SUPERINFECTION)

Age in days (t)	Incidence data (h)	Prevalence data (d)	("r")
12	0.00878	0.006043	0.002737
42	0.008486	0.006536	0.00195
72	0.008814	0.006416	0.002398
102	0.009356	0.006317	0.003039
132	0.009917	0.006040	0.003877
162	0.01024	0.005729	0.004511
192	0.01024	0.00525	0.00499

$$h = \frac{-\log x}{0.4343 t}$$

Explanation of symbols employed in this section

- m = anopheline density in relation to man.
- a = average number of men bitten by one mosquito in a day.
- b = proportion of inocula which cause an infection.
- p = the probability of an anopheline surviving through one day.
- n = time taken for completion of an extrinsic cycle.
- s = proportion of mosquitos with sporozoites in the salivary glands.
- h = proportion of population receiving infective inocula in one day.
- x = proportion of population infected.
- r = proportion of infected people who revert to the uninfected state in one day.
- t = time in days.
- e = the base of natural logarithms: 2.71828.
- d = apparent inoculation rate, derived from prevalence data.
- HB = human blood.

TABLE 13. INOCULATION RATE OF A. GAMBIAE (EVALUATION AREA)

Month and year	Sporozoites rate % (1)	Biting rate per bait night (2)	Fed PSC per man night (3)	Fed ETC per man night (4)	Total fed per man night (5)	Total HB fed per man night (6)	Inoculation rate (h) Biting (7)	Inoculation rate (h) Fed (8)
	s	ma			f	f x 0.95	mas	s x (6)
1972								
January	4.18	-	1.54	-				
February	4.36	8.33	2.07	0.03	2.10	2.00	0.363	0.087
March	8.97	3.07	2.13	0.02	2.15	2.04	0.275	0.183
April	3.65	4.56	3.71	0.02	3.73	3.54	0.166	0.129
May	1.58	32.79	10.39	0.13	10.52	9.99	0.518	0.158
June	3.64	30.83	10.76	0.13	10.89	10.35	1.122	0.377
July	6.91	10.54	5.08	0.05	5.13	4.87	0.728	0.337
August	9.11	1.47	1.89	0.00	1.89	1.80	0.134	0.164
September	6.36	2.29	1.02	0.02	1.04	0.99	0.146	0.063
October	6.23	0.50	0.79	0.00	0.79	0.75	0.031	0.047
November	1.18	12.83	5.48	0.03	5.51	5.23	0.151	0.062
December	5.31	21.33	8.49	0.02	8.51	8.08	1.133	0.429
1973								
January	10.60	5.70	1.92	0.00	1.92	1.82	0.604	0.193
February	8.86	3.14	1.81	0.00	1.81	1.72	0.278	0.152
March	11.34	1.64	1.20	0.00	1.20	1.14	0.186	0.129
April	6.07	2.67	1.94	0.00	1.94	1.84	0.162	0.112
May	2.40	15.40	7.21	0.02	7.23	6.87	0.370	0.165
June	4.80	8.33	6.72	0.08	6.80	6.46	0.400	0.310
July	13.16	2.07	1.61	0.01	1.62	1.54	0.272	0.203
Total	5.86	9.65	3.91	0.04	3.95	3.75	0.565	0.220

TABLE 14. INOCULATION RATE OF A. FUNESTUS
(EVALUATION AREA)

Month and year	Sporozoites rate % (1)	Biting rate per bait night (2)	Fed PSC per man night (3)	Fed ETC per man night (4)	Total fed per man night (5)	Total HB fed per man night (6)	Inoculation rate (h) Biting (7)	Inoculation rate (h) Fed (8)
	s	ma			f	f x 0.99	mas	s x (6)
1972								
January	1.99	-	0.96	-				
February	7.73	3.39	2.03	0.00	2.03	2.01	0.262	0.155
March	6.52	4.14	2.17	0.02	2.19	2.17	0.270	0.141
April	5.89	5.81	2.01	0.01	2.02	2.00	0.342	0.118
May	2.76	12.48	3.06	0.01	3.07	3.04	0.344	0.084
June	1.97	10.88	5.12	0.06	5.18	5.13	0.214	0.101
July	3.80	8.79	4.81	0.22	5.03	4.98	0.334	0.189
August	5.00	2.50	3.00	0.02	3.02	2.99	0.125	0.150
September	4.20	6.63	2.20	0.02	2.22	2.20	0.278	0.092
October	6.93	1.97	1.64	0.00	1.64	1.62	0.137	0.112
November	2.23	3.77	2.63	0.00	2.63	2.60	0.084	0.058
December	2.99	5.25	4.20	0.02	4.22	4.18	0.157	0.125
1973								
January	5.81	10.50	2.18	0.03	2.21	2.19	0.610	0.127
February	5.09	7.94	3.46	0.01	3.47	3.44	0.404	0.175
March	6.80	5.75	3.46	0.01	3.47	3.44	0.391	0.234
April	3.95	3.81	4.66	0.03	4.69	4.64	0.150	0.183
May	3.11	8.13	4.19	0.02	4.21	4.17	0.253	0.130
June	1.42	7.06	9.86	0.04	9.90	9.80	0.100	0.139
July	7.50	4.31	4.36	0.01	4.37	4.33	0.323	0.325
Total	4.41	6.53	3.34	0.03	3.37	3.34	0.288	0.147

TABLE 15. COMBINED INOCULATION RATES OF
A. GAMBIAE AND A. FUNESTUS (EVALUATION AREA)

Month and year	Combined inoculation rate (h) Biting	Combined inoculation rate (h) Fed
1972		
January		
February	0.625	0.242
March	0.545	0.324
April	0.508	0.247
May	0.862	0.242
June	1.336	0.478
July	1.062	0.526
August	0.259	0.314
September	0.424	0.155
October	0.168	0.159
November	0.235	0.120
December	1.290	0.554
1973		
January	1.214	0.320
February	0.682	0.327
March	0.577	0.363
April	0.312	0.295
May	0.623	0.295
June	0.500	0.449
July	0.595	0.528
Total	0.853	0.367

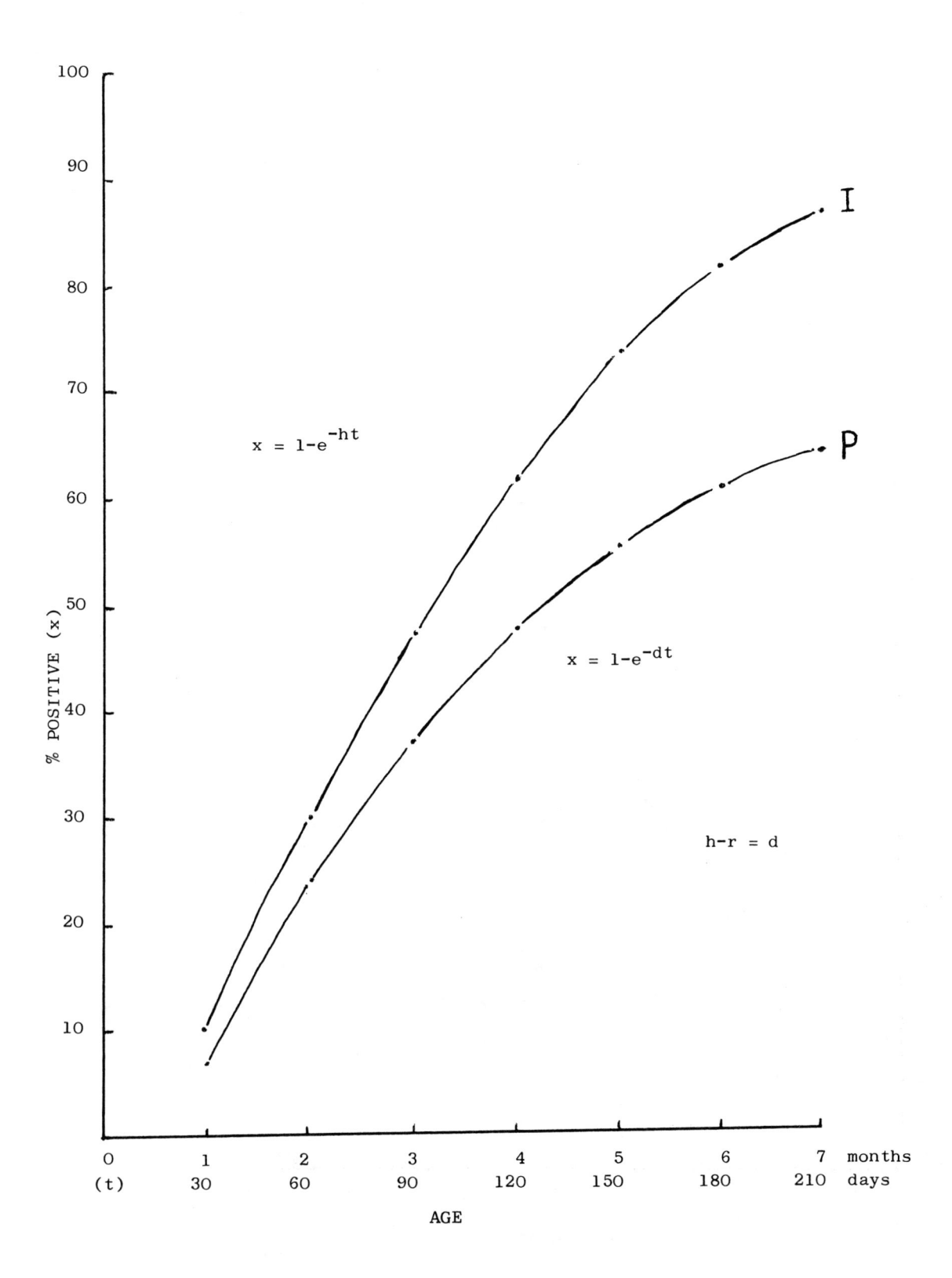

Fig. 22. Age curves of disease acquisition (I - derived from incidence data: P - derived from prevalence data).

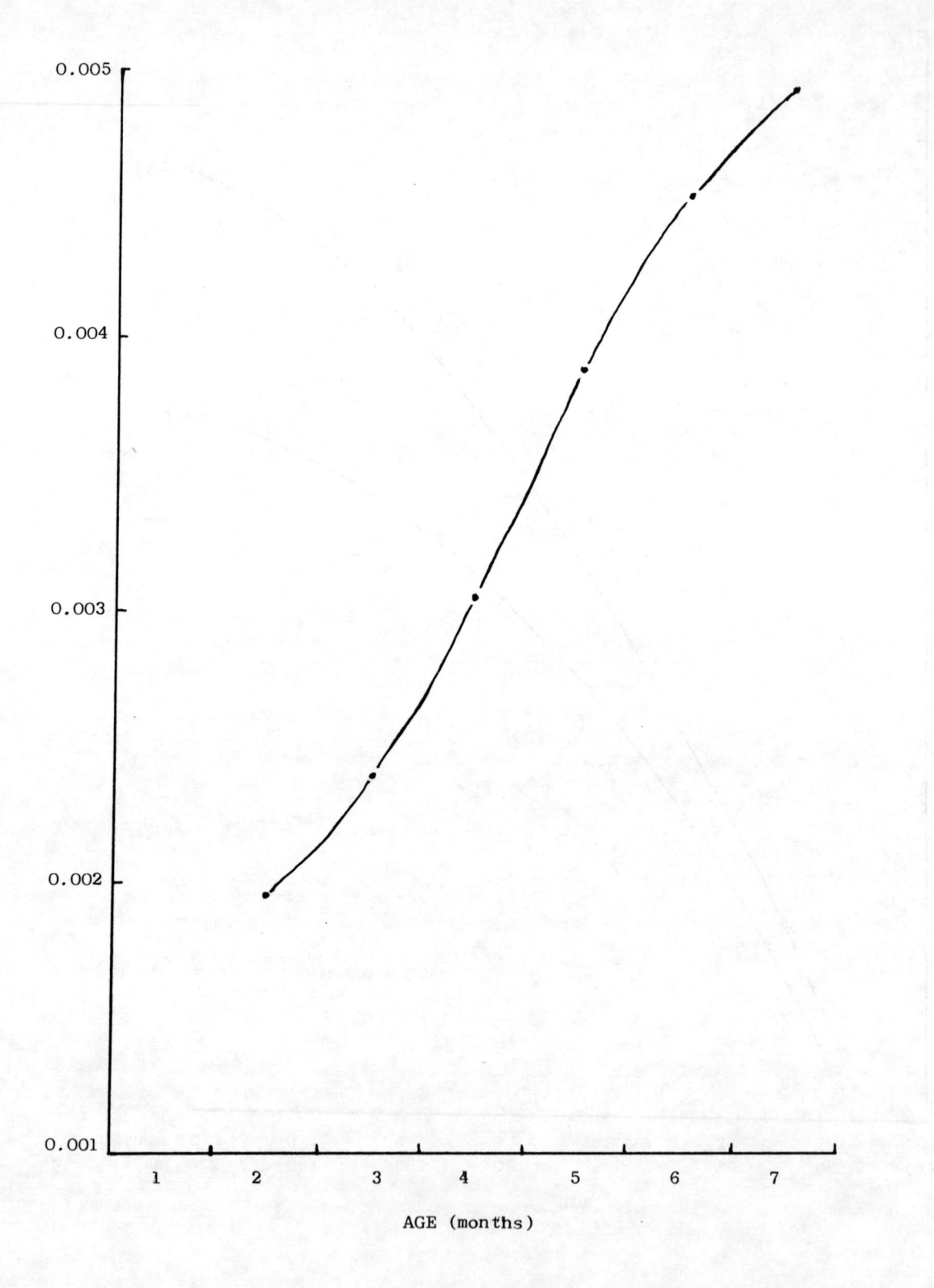

Fig. 23. The value of "r" as derived from Fig. 22, that is ($\underline{h}-\underline{d}$), or actual $\underline{r}$ modified by superinfection.

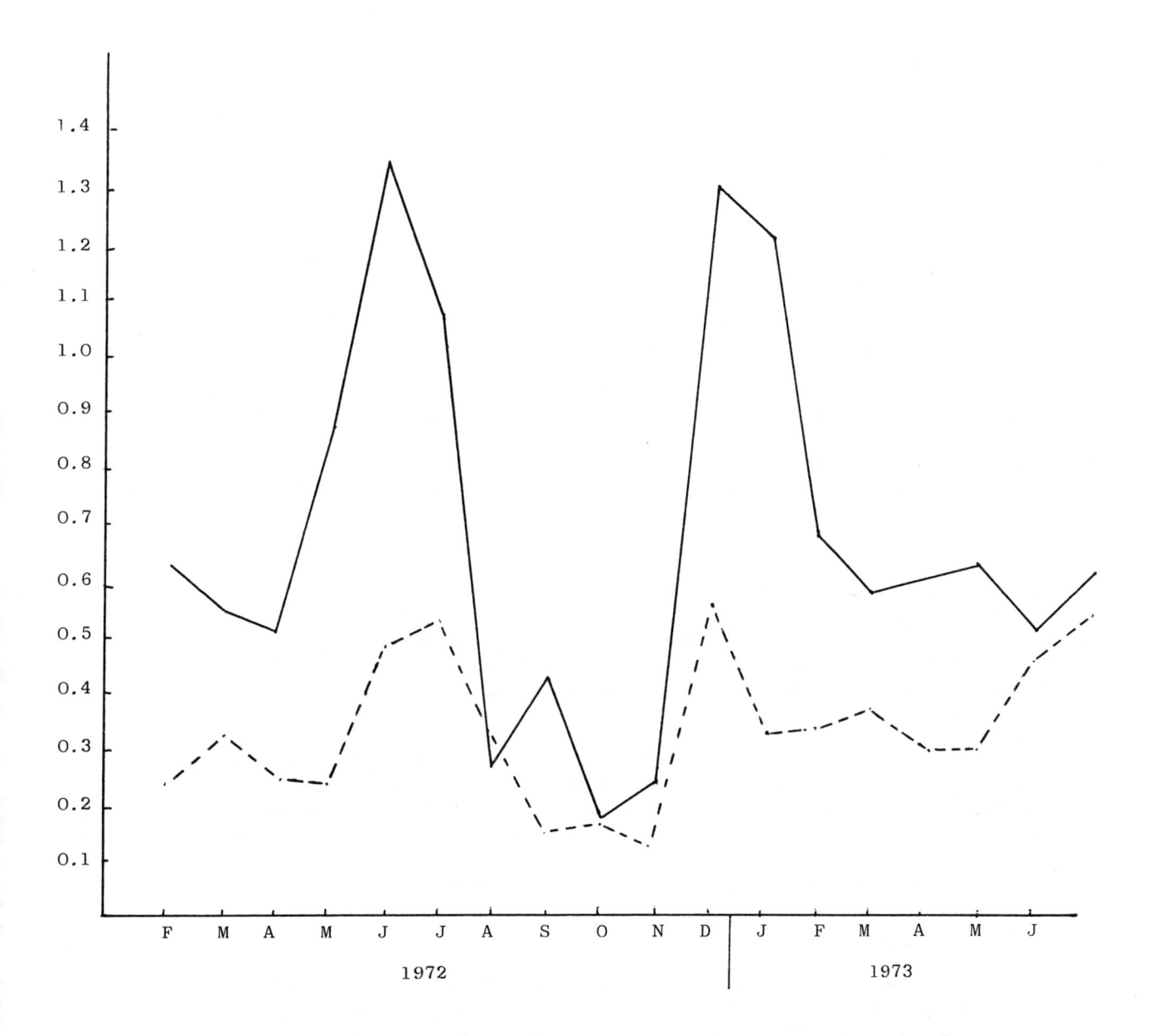

Fig. 24. Daily inoculation rate for A. gambiae plus A. funestus (evaluation area): solid line = man-biting sample, dashes = pyrethrum spray and exit trap samples (blood fed).

The values obtained were:

for man-biting sample: $\underline{h}$ = 0.853
for pyrethrum sample, and exit trap collection: $\underline{h}$ = 0.367

The difference between these two rates may be at least partially, if not entirely, explained by the particular technique employed in man-biting catches which may have increased the attractiveness of the huts used above the normal, resulting in an inflated biting rate. It is thus likely that the actual value of the entomological daily inoculation rate lies between the above two figures, and nearer to 0.367 than to 0.853; the value of 0.5 may be taken to be a close approximation.

In calculating the above data it will be seen that there is a considerable difference between the inoculation rates estimated on the basis of entomological data on the one hand and parasitological (infant) data on the other. This difference may be taken as a measure of the factor $\underline{b}$ (the proportion of infected bites which actually cause an infection).

Thus, for the infant data above, $\underline{b}$ would be in the region of $\frac{0.0083}{0.5}$ = 0.0166. That is, out of every 100 infected bites only 1.6, on the average, would cause infections (considering infants as exposed to the same biting rate as adults) (Macdonald, 1973). Since the daily entomological inoculation rate is an average figure it may be reasonably asserted that the factor $\underline{b}$ does not affect each person equally, its average value being probably the result of various variables. Among the latter, it is likely that infants are bitten less frequently than adults due to their smaller size and protection by clothes. They are also under the influence of maternal immunity during the first month of life. This is borne out by the distribution of primo-conversions from negative to positive according to age in months (Fig. 22).

Of comparable epidemiological importance to the anopheline daily survival rate is the time required for completion of the extrinsic cycle of the parasite. To gain some idea of this, one may apply the method of Moshkovsky (1946) to estimate the time required for completion of sporogony by the two main species. This involves the "sum of heat" theory according to which a certain total of degree-days (centigrade) above the threshold temperature is required for completion of the extrinsic cycle. The threshold temperature for extrinsic development of both $\underline{P.\ falciparum}$ and $\underline{P.\ malariae}$ is taken as 16°C. The heat requirements mentioned above are 111 degree-days for $\underline{P.\ falciparum}$ and 144 degree-days for $\underline{P.\ malariae}$. As the average mean daily temperature in the trial area during the observation period was 23.2°C, this means that the average length of the extrinsic cycle of $\underline{P.\ falciparum}$ was $\frac{111}{7.2}$ = 15.4 days, and that of $\underline{P.\ malariae}$ $\frac{144}{7.2}$ = 20 days, thus giving a difference of 4.6 days in the time required for sporogony.

2.1.2 General comment

No attempt has been made to define malaria according to existing classifications of endemicities as they do not take into account sufficiently the differences which do in fact exist between areas of apparently similar levels of endemicity. For example, in classifications of endemicity commonly used, no reference is made to the most valuable index of parasite count or to the differences in the chances of acquiring malaria infection. Malaria should be classified according to type, account being taken of the causative malaria parasite, the anopheline infestation and infectivity for man, the chances of man acquiring the disease, including superinfection, human susceptibility and reaction to malaria. At present, the available data are too incomplete to draw any conclusions for meaningful classification of the malaria type, taking into account all the main epidemiological parameters. However, regarding malaria in Kisumu area (Kenya), it can be stated that malaria is characterized by the following aspects:

   (i)   parasite rates and parasite load are at their highest in the age group 1-4 and decrease steadily with age;

   (ii) in adults above 15 years of age, the mean malaria parasite load is only a small fraction of that seen in the age group 1-4 years;

   (iii) there is little seasonal rise in human parasitization except in infants;

   (iv) the malaria parasite species involved are mainly P. falciparum and P. malariae, with a marked predominance of the former;

   (v)   the vectors A. gambiae and A. funestus are present all year round and are considered as mostly anthropophilic and endophilic.

2.2. EXAMPLES OF INTERPRETATION OF DATA OBTAINED FROM AREAS UNDER ATTACK MEASURES

2.2.1 A case of vector resistance to insecticide

The important example is the case of insecticide resistance in A. stephensi in some countries in the Eastern Mediterranean Region. DDT resistance was first reported from Saudi Arabia in 1955. In Iran and Iraq, A. stephensi resistance to DDT was found to be widespread in 1957 shortly after the malaria eradication programme started which eventually necessitated replacing DDT by dieldrin in 1959. Dieldrin resistance appeared after two years of its application and a malaria outbreak occurred when A. stephensi density reached a very high level.

A decrease in the DDT resistance level and the favourable effect obtained when DDT was applied for combating the malaria outbreak encouraged continuation of its use. It was applied in 2 rounds annually at a rate of $2g/m^2$ in each round, commencing 1962-63 in the Khuzistan plains in Iran and in Basrah Liwa in Iraq and it was supplemented by other measures in certain areas.

In October 1967, in Iran, the second annual spraying of DDT was replaced by malathion at $2g/m^2$ in order to assess its efficacy in controlling the autumn transmission. Continuous epidemiological/entomological evaluation during 1965-68 showed that transmission continued to persist in both areas. Applying different entomological techniques gave a good support to the parasitological findings which showed little or no response to DDT applied in two annual spraying rounds as presented graphically in Fig. 25. The build-up of the seasonal density of A. stephensi continued together with a high parous rate indicating a high probability of daily survival (Table 16). This was also confirmed by finding a high survival (93%) in window-trap collections held for 24 hours. At the same time there had been a continuous fall in the susceptibility level of A. stephensi, the mortality on the standard exposure to DDT, which was about 56% in 1961, reaching 37% in 1965, 30% in 1966, 21% in 1967 and 16% in July 1968 after the spring spraying round of DDT.

TABLE 16. THE ENTOMOLOGICAL INDICES AS OBTAINED BY
DIFFERENT TECHNIQUES ON A. STEPHENSI

Date	% Parous	Estimated probability of survival through one day (P)[a]	% Survival in window-trap collections	% Survival on 4% DDT for 1 hour
1965				
August	67.5	0.82	93 (21)	-
September	69.1	0.83	92 (124)	63
1966				
August	57.8	0.76	97.6(39)	-
September	78.7	0.887	93 (112)	-
October (after round II, DDT)	49.7	0.705	-	70
1967				
August	68.8	0.83	93.3(15)	-
September	79.3	0.89	94 (73)	-
1968				
July	55.4	0.744	-[b]	84
August	57.5	0.76	-[b]	-

[a] Assuming that the duration of the gonotrophic cycle throughout the season is two days.

[b] In 1968 no A. stephensi could be obtained by bait capture or by window-trap collections.

A malathion spraying round applied in September/October 1967 seemed to have much reduced the density of A. stephensi and the parasite incidence, but since the seasonal decline intervened no firm conclusion could be drawn. In 1968, the spring spraying was again with DDT and the second round was with malathion being advanced to commence in mid-August. The entomological observations showed that after the DDT round in the spring, A. stephensi density exhibited a rising trend but at a much lower level than in the previous years.

Vector longevity as determined from the parous rate, during the post-operational period of DDT spraying was relatively lower than that observed in previous years under 2 annual spraying rounds of DDT. The parasitological trends agreed well with slowly developing vector density and longevity in 1968. Malathion applied in August drastically reduced the population of A. stephensi. On the basis of ample parasitological/entomological evidence, it appeared that there was no advantage of DDT spring spraying since, because of the stepping up of the level of DDT resistance in A. stephensi, it is possible that achievement of interruption of transmission would take longer than required. Therefore, a timely decision was taken to replace DDT by malathion. Thus 2 rounds of malathion spraying have been applied annually in the area where A. stephensi is the only vector. Where, in addition to A. stephensi, other vectors exhibiting exophilic tendencies occur, as in the "foothill areas", and where the problem of nomadism also exists, the DDT/malathion regimen was maintained but supplemented by mass drug administration. At the same time surveillance was reinforced to cover the whole population, chemical larviciding was carried out in and around towns and cities, and the use of Gambusia was extended.

With these measures favourable results have been achieved. The reduction in vector density has been notable and a decrease in the parasite rate of 30-90% was observed, the greatest decrease being in the areas where A. stephensi was the only vector.

Similar results were obtained in Iraq after a timely decision was taken to replace DDT with malathion in the southern region as from 1969. The results show that transmission has been interrupted in the problem area where DDT/dieldrin-resistant A. stephensi is the vector.

2.2.2 A case of a vector exhibiting an exophilic tendency and becoming resistant to insecticide.

In 1959, DDT tolerance and partial exophily were observed in A. pharoenis in Egypt. Hence it was decided to assess the effectiveness of DDT house spraying to be applied in one round per annum at a dosage of $2g/m^2$ commencing in 1960.

The application of the annual spraying round was completed before the onset of the transmission season which usually starts about the beginning of July. The trial area was situated in the southern rural area of the Nile delta. The area was about 58 km^2 with a population of 25 000, whose main occupation was mixed agriculture, growing cotton, maize and vegetables. Rice cultivation was patchy. The man:animal ratio was 1:7. The principal vector of malaria was A. pharoensis.

Baseline data, 1959

Prespraying observations showed that A. pharoensis was very selective in choosing the day-time indoor resting shelters, as it was found to concentrate in high density in a few suitable rooms. Large densities were observed in animal rooms.

- Collections by spray capture from houses were characterized by a higher proportion of bloodfed than gravid females. This gave the first indication that A. pharoensis exhibits an exophilic tendency, at least passing part of the gonotrophic cycle outdoors.

- This was confirmed by observations on its exodus from houses by means of outlet window-traps and by finding it resting during the day-time in plants at different stages of the gonotrophic cycle.

- Very high biting densities were recorded on man and animals indoors and outdoors.

- The overall human blood index determined in samples of bloodmeal smears taken from indoor and outdoor shelters was about 31%.

- The average duration of the gonotrophic cycle was estimated to be 2 days.

- Exposure to 4% DDT for 1 hour gave 81-90% mortality. The LC_{50} was 1.8-27%.

Post-spraying evaluation

Entomological observations were pursued throughout 1960-62 in the sprayed and unsprayed areas, utilizing the following techniques:

- window traps, for determining the trap kill,
- age-grouping techniques, for determining the proportion parous,
- periodical bait capture, for estimating the man/vector contact,
- susceptibility testing, for determining the changes in the susceptibility level of A. pharoensis.

Parasitological surveys were simultaneously conducted, and infants born after spraying were subjected to repeated blood examinations.

Fig. 26 shows the trap density and mortality, and the percentage mortality with 1 hour exposure to 4% DDT for qualitative assessment. Fig. 27 shows the results of parasitological surveys expressed as expected and actual rates of decline in the parasite rate of the general population, using the method of Macdonald & Göckel (1964).

For quantitative assessment the following formulae of Macdonald (1957) were utilized:

- p: the probability of daily survival (the square root of the proportion parous on the basis of the gonotrophic cycle being 2 days).

- p^n: the probability of survival through the duration of the sporogonic cycle. (The period of sporogony (n) has been estimated for Plasmodium vivax according to the mean temperature during the different periods of observations, from Oganov-Rayevski Tables - in Detinova, 1962).

(a)   Assessment of 1960 spraying (Fig. 26)

(i)   Three weeks after spraying in June, the trap mortality was 96%, but it dropped to 64% after 8 weeks, and to 57% after 13 weeks. The average trap mortality was 70% during the first 8 weeks.

(ii)   The parous rate was markedly reduced shortly after spraying but gradually increased as the effect of the DDT became weaker, and because of some operational defects. The parous rate in the thirteenth week in the sprayed area was 54%, and in the unsprayed area was 57% in the biting samples.

(iii) Exposure to 4% DDT for 1 hour gave 58-94% mortality and the LC_{50} reached about 3.8% indicating a slight deterioration in the susceptibility level of A. pharoensis in the first year of DDT spraying.

(iv)   The overall parasite rate (Fig. 27) declined to 20% of that of one year earlier, indicating a satisfactory rate of fall in the first year of attack, with the exception of a temporary reverse during September when 1.2% of the infants examined were found positive, this finding reflecting operational defects together with some loss of insecticide. This rate of fall, if maintained, would ultimately have resulted in diminution of malaria to a very low level.

(v)   The probability of daily survival (p) was estimated to be 0.5 in the vector population as a whole, and slightly higher 0.55 in the biting population. The probability of survival through the sporogonic cycle (p^n) was approximately 0.002.

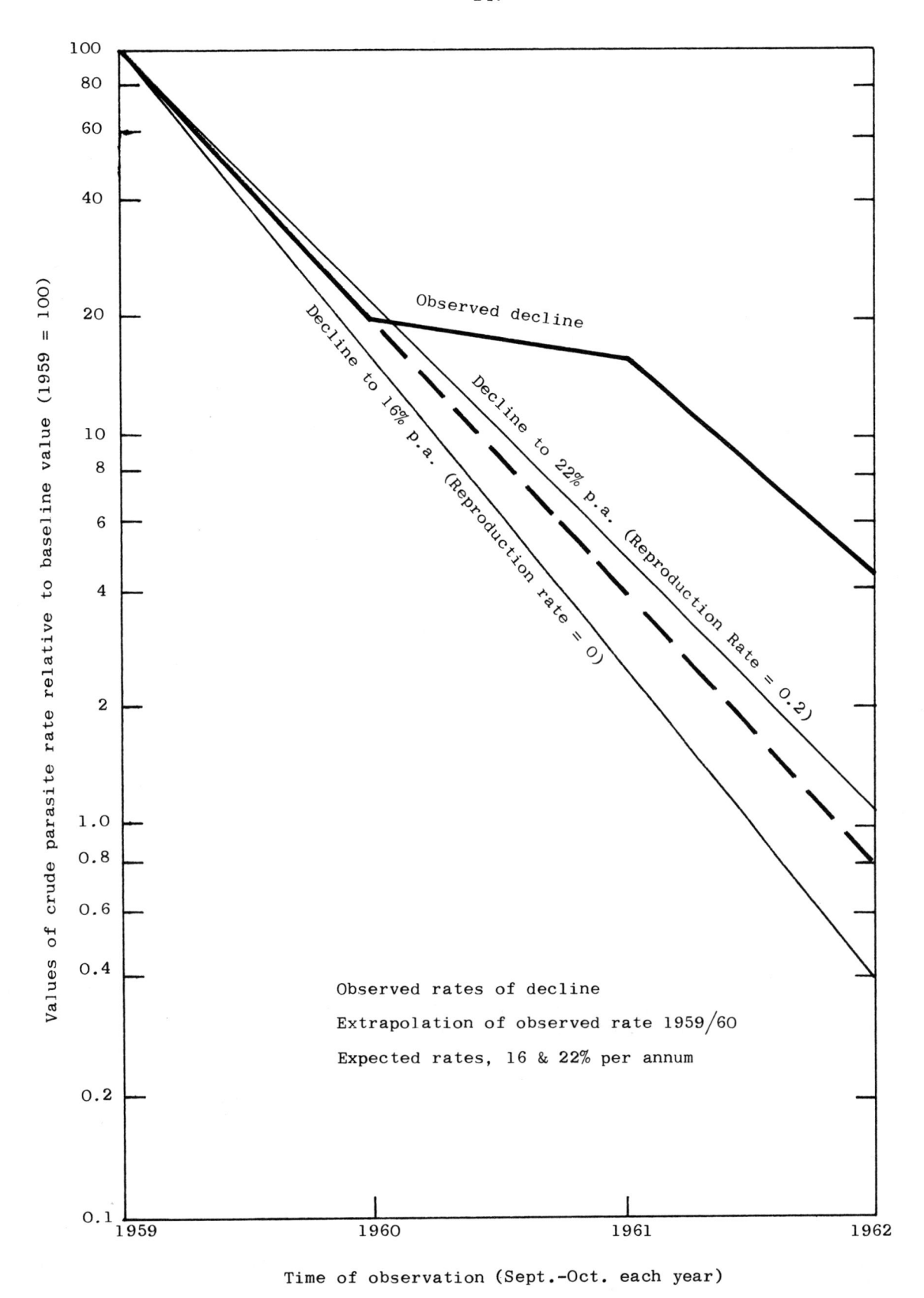

Fig. 27. Expected and actual rates of decline in the parasite rate of the general population (1959 = 100)

As the parasitological picture was on the whole satisfactory, the above parameters were taken as a standard for the critical value above which transmission would continue. At the same time the trap mortality of 70% or more indicated satisfactory control: this value is regarded as unsatisfactory when it falls to 54%. Against this standard of critical value, the DDT spraying of the subsequent years was assessed.

(b) Assessment of 1961

DDT spraying was completed by the first week of June. The following results were obtained:

(i) High trap mortality, 81-65%, was obtained for about 8 weeks after spraying, but it dropped to 55% in the 10th and 11th weeks. From the 12th week to the 17th week mortality fluctuated between 37% and 20%. The overall trap mortality was slightly over half of the 70% level required for satisfactory control.

(ii) Vector probability of daily survival was 0.62 and the probability of survival through the sporogonic cycle 0.013, over 6 times the standard suggested. Neither index reached that level in the unsprayed area.

(iii) Further deterioration in the susceptibility level of A. pharoensis was indicated from the mortality of 7-66% obtained with one hour exposure to 4% DDT.

(iv) The crude parasite rate showed a much slower rate of decline as it fell only to 16% of the baseline instead of to 4% of it as expected after two years of attack, had the initial rate of decline been maintained.

(c) Assessment for 1962

DDT spraying was intentionally delayed until the second week of July, in the hope that its effectiveness would cover the critical period of the transmission season, August-October. The following results were obtained:

(i) Trap mortality was around 50% for only about 4 weeks, followed by a drop to less than 20% in September. Thus for a long period in the transmission season the population remained unprotected.

(ii) The critical level of vector probability of daily survival was slightly exceeded, but that for the probability of survival through the sporogonic cycle (0.012) was exceeded by a bigger margin.

(iii) The drop in effectiveness of DDT in 1962 was also correlated with continued deterioration in the susceptibility level of A. pharoensis as the level of mortality on 4% DDT for 1 hour was 0-20%.

(iv) The general parasite rate fell to about 28% of the value of 1961, i.e., 4.5% of the original value of 1959. This may be compared with the reduction to 0.4% of the 1959 level expected in 1961 had satisfactory control been continuously maintained. The slope of the curve implies further low-grade transmission in the sprayed area in 1962.

Conclusion

From the 1961 and 1962 data, the unsatisfactory control coincided with deterioration in the susceptibility level of A. pharoensis to DDT. However, it could not be ascertained whether this was the main responsible factor, or if an important role was played by outdoor biting and resting of A. pharoensis population.

Progressive selection for DDT resistance was also observed in unsprayed areas due to high selection pressure of insecticides extensively used in agriculture commencing 1961, for combating a voracious attack of cotton leaf worm.

Annex 1

OUTLINE FOR ANALYSIS, INTERPRETATION AND PRESENTATION OF ENTOMOLOGICAL AND RELATED DATA

PRELIMINARY SURVEYS IN AREAS DESIGNATED FOR ANTI-VECTOR ACTIVITIES

The preliminary survey includes reconnaissance of the area and is a first step towards selection of the appropriate techniques and collection of baseline information. It will lead to the selection of indicator villages. The report of such a survey should:

(a) Indicate the area covered by the entomological survey and how it is related to the area covered by the malariometric survey. Define the suitability of the seasons in which each of the two types of survey was carried out.

(b) List the anopheline fauna in each ecological substratum from the existing information and those species which were or were not encountered plus any additional species found during the survey.

(c) Define the recognized vector(s) detected and sporozoite rate if determined at the time of the survey, and give the geographical extent of vector dominance.

(d) Indicate availability of breeding places, and the influence of meteorological conditions from the existing information and as recorded during the survey.

(e) Give the following entomological findings:

- the density of vectors in the various types of house together with a detailed description of the types of house found with high density;

- the resting sites within premises for each vector and the maximum height of such sites;

- the density in houses as compared with biting density, from night observations carried out simultaneously in certain localities;

- degree of man biting by vectors indoors and outdoors, and the relation of this to human activities at night and to sleeping habits;

- degree of man and animal biting and the relation of this to the human:animal ratio in the area in comparison with any results of precipitin tests made;

- degree of exophily as indicated from the abdominal stages of house-resting mosquitos and any found in trap collections.

(f) Give the susceptibility levels of vectors as determined on adequate samples of vectors with a view to the initial choice of the insecticide to be used. Give any information that could be collected on the history of insecticidal treatment, including agricultural pest control.

(g) Define the following epidemiological factors:

- the correlation between the meteorological conditions and the density and infectivity of vectors during the period of the survey;

Annex 1

- the possible role of each vector in malaria transmission and its relation to malaria endemicity and stability in the area surveyed.

(h) Combine the malariometric/entomological information on a map, showing vector distribution and endemicity levels.

(i) State whether the entomological observations were satisfactory in quantity, quality and timing.

TREND OBSERVATION AT FIXED INDICATOR VILLAGES FOR BASE-LINE DATA

Interpretation of the trend observations should:

(1) Define the resting density in houses vis-a-vis the man biting density indoors and outdoors, and relate these to human activities at night, and sleeping habits; and indicate the man and animal biting with reference to human:cattle:animal ratios. Precipitin results should be related to man and animal biting.

(2) Indicate the trend of density and infectivity of vector populations resting in houses and show if trends in infectivity are in agreement with those of the biting population.

(3) Consider the results of continued observations on the resting sites of vector species within premises and the maximum height of such sites. This should also include observations on various unusual resting sites appearing under certain conditions, such as granaries, chicken pens, etc.

(4) Indicate whether there is evidence of an exophilic tendency as may have been deduced from the abdominal stages of day-time house-resting mosquitos, and the consistent presence of a significant proportion of blood-fed females in window traps. Indicate if this was confirmed by finding blood-fed and gravid vector females in outside shelters (whether natural or artificial) and including the underside of raised floors of houses as may occur in some areas.

(5) Indicate the significance of such an outside resting population, based on the presence of specimens with positive reaction for human blood and any results of dissection for infectivity and parity.

(6) Demonstrate the trend of parous rate, if the respective techniques are applied on samples taken from all situations. When analysing this trend, full consideration should be given to the prevailing density and the size of samples of vectors dissected.

(7) Indicate any discrepancy in entomological trends and discuss the possible reasons, i.e., whether they are due to faults in techniques or to other factors involved, and the action intended to correct such discrepancy in the future.

(8) Give the susceptibility levels of vectors with a view to the choice of insecticide to be used in the campaign and establish the discriminating dosage to be used for periodical checking. Give further information collected on the history of insecticide treatment and discuss the possible implications of any agricultural pest control underway or planned.

(9) Discuss the following epidemiological factors:

(a) the trends of density, infectivity and parity (preferably estimate the probability of each vector's survival through one day and through the extrinsic cycle in different

Annex 1

seasons) in comparison with the trend of the parasitological indices, correlating these with meteorological conditions; classify the vectors as principal and secondary in relation to their role in malaria transmission;

(b) the combined role of all vectors, defining the transmission season in comparison with the parasitological findings;

(c) any discrepancy between the entomological and parasitological findings and give explanation.

(10) Indicate whether the entomological observations were satisfactory in quantity, quality and timing.

The above should be correlated throughout with meteorological conditions.

TREND OBSERVATIONS AT FIXED INDICATOR VILLAGES IN AREAS UNDER RESIDUAL HOUSE SPRAYING

It is important to define the period elapsed from the date of spraying of each indicator village and give the available evidence that spraying coverage and quality met the required standards, since the aim of trend observations is to assess the impact of good spraying. The interpretation of data should:

(1) Discuss the general trend of vector(s) densities as compared with pre-spraying baseline data as drawn from:

- daytime resting in houses as compared with density of man-biting indoors and outdoors,

- window trap collections as compared with day-time indoor and outdoor resting.

The best approach is to follow these trends graphically together with the parasitological trends.

(2) Involve further analyses to see:

(a) if the overall reduction in density of vectors as observed by all the above sampling techniques is also influenced by meteorological conditions at certain seasons;

(b) if vectors have been absent or scarce in sprayed premises during day-time searches although found by bait capture indoors and outdoors; if so, compare the density and the proportions of the abdominal stages of trap collections before and after spraying. Give the trap mortality throughout the period of observations and correlate it with density in traps and those of other captures.

From this analysis, indicate whether one or more of the following conditions are met for the principal and secondary vectors:-

- that vectors enter sprayed premises, feed and escape unharmed.

- that there is an increased exodus of the unfed, and that this was not due to the fact that the baits covered themselves completely, or to any other factor such as disturbance by smoke.

- that there is a high survival rate in window trap collections and high biting densities due to deterioration of effectiveness of the insecticide in relation to the time lapse since the spraying round has been applied or due to habits of the inhabitants, e.g. replastering, rethatching, etc.

Annex 1

- that the principal vector is becoming resistant to the insecticide but not to the extent that large numbers as yet rest in sprayed houses, in which case give the results according to the classification recommended in Part I, Chapter 1, section 7. Indicate whether the presence of insecticide resistance has been verified and confirmed. List the insecticides used in agriculture.

- that as indicated by the presence of high man-biting densities but low trap density, vectors scarcely enter sprayed premises, but largely feed outdoors on man and rest in outside shelters (relate this to sleeping habits and movements of people at night to illustrate the chances of outdoor transmission, correlating the above behaviour with the trend of density of the outside resting mosquito population, particularly that of the fed and half gravid proportions. Illustrate the significance of this population by precipitin tests and gland dissection).

- that otherwise a shift has been made in the feeding habits, related to influence of animal population in the area.

(3) Give the probability of survival as estimated from the trend of parous rate determined on samples drawn from man-biting catches with full consideration to the level of densities hitherto encountered. Compare this estimated survival of vector(s) and the survival observed in window trap collection and in susceptibility testing.

(4) Indicate any discrepancy in entomological trends, explaining whether this was due to faults in techniques or any other factors, and what action will be taken to correct this in future.

(5) On the epidemiological side:

(a) indicate whether the favourable entomological trends are supported by the parasitological trends;

(b) otherwise indicate whether the findings of entomological evidence of breakdown in vector control was followed by parasitological evidence that transmission is continuing or whether there is any discrepancy between the two trends, and give explanations;

(c) if transmission is continuing in the area represented by indicator villages, indicate whether this is due to:

- insufficiency of the insecticide dosage or faulty timing and frequency of application of spraying rounds (particularly in initial years of attack).

- inadequate maintenance of spraying coverage to cope with human interference and the construction of temporary huts between spraying rounds.

- one of the vectors not responding to attack measures or another vector being suspected (particularly in prolonged period of attack).

- vector resistance being responsible (particularly in prolonged attack).

- vector outdoor biting and exophilic behaviour coupled with human activities at night being responsible for outdoor transmission.

Annex 1

SPOT CHECKS

In areas under residual house spraying interpretation of results should:

(1) Give the distribution of localities surveyed and give the reasons for the selection of these localities and the date of spraying of each locality.

(2) Discuss the condition of spraying in houses surveyed and indicate where the vector(s) was found in high density.

(3) Compare the vector house-resting density in sprayed premises, disturbed and unsprayed premises with the density levels obtained from trend observations at the indicator villages.

(4) Discuss the blood digestion stages of collections obtained from each of the above premises, and whether gravid females were present in sprayed premises, giving reasons whether this is due to deterioration of effectiveness of the insecticides or that resistance is suspected and whether susceptibility checking has been done (if so give results). In the case of an exophilic vector, give the man-biting density in and around unsprayed and newly constructed and replastered houses. Compare these indices with those obtained from the fixed indicator villages.

(5) Indicate the extent of reduction in spraying coverage, such as structural changes in sprayed premises, missed houses, new constructions, and temporary summer huts remaining unsprayed.

(6) Show the extent of the area and population which has been affected by operational defects, if possible.

(7) Show parasitological findings obtained from localities discovered to have been affected by operational defects, and where entomological evidence has already been provided. Indicate remedial action undertaken and the results of entomological follow-up.

In areas under consolidation interpretation of results should:

(1) Give the distribution of localities surveyed and give the reasons for the selection of these localities.

(2) Give the date of last spraying.

(3) Give the environmental conditions prevailing during the survey.

(4) Indicate whether all vectors are still present and whether they have regained their original density, including:

(a) indoor resting density for endophilic species, and

(b) man-biting density indoors and outdoors for exophilic vectors.

(5) Indicate whether there is a need for further entomological/parasitological investigation in view of the vulnerability of the area and/or man-made development favouring an increase in vector output.

Annex 1

FOCI INVESTIGATIONS

In areas under residual house spraying interpretation of data should:

(1) Give the area affected by malaria transmission and distribution of localities of indigenous cases.

(2) Give the localities of foci selected for entomological investigation.

(3) Define for each focus (preferably on a chart as in Fig. 6),

- the number of indigenous cases and Plasmodium species and gametocyte carriers,

- the date of their detection and dates of onset of the primary attack as revealed from the epidemiological investigation,

- the date of spraying in relation to the date of onset of the primary attack,

- the prevailing meteorological conditions,

- the date of the entomological investigation, as this has important bearing on interpretation of data.

(4) Indicate whether some indigenous cases were re-investigated epidemiologically and give results of fever surveys of collaterals.

(5) Enumerate the techniques employed as guided by those utilized for trend studies and any additional techniques and the reasons for their choice.

(6) Give the quality of spraying for the locality of the focus as whole and for houses of the indigenous cases and surrounding houses.

(7) Give conclusions as to the possible causes of continuation of transmission whether due to:

(a) operational factors, such as:

- deterioration of effectiveness of the insecticide due to faulty dosage, faulty timing of application, local climatic conditions or large-scale human interference.

- new, missed seasonal huts remaining unsprayed.

(b) biological factors such as:

- increased exodus of vectors from sprayed houses unharmed and the use of outdoor day-time shelters

- a proportion of the human population sleeping or remaining active outdoors at night

- vector resistance to the insecticide used.

(c) any combination of the above.

Annex 1

(8) Suggest remedial action for effective control measures on vector(s) involved in the light of causes shown in (7) above.

(9) Describe any follow-up observations for assessing the remedial action.

In areas where residual house spraying has been withdrawn - (e.g., consolidation, maintenance and where malaria eradication has been certified) interpretation of the results of observations should:

(1) Describe the localities affected by recurrence of malaria transmission and their distribution and show any important environmental changes.

(2) Give the date on which the spraying was discontinued.

(3) Give the localities selected for entomological investigation.

(4) Give the epidemiological evidence for the origin and type of infection and the number and distribution of secondary cases.

(5) Give the time lapse between detection of the cases and the epidemiological/entomological investigation and the prevailing climatic conditions as this has important bearings on interpretation of data.

(6) Enumerate the techniques employed, as guided by past experience during the application of attack measures. Indicate whether any additional techniques have been utilized and the reasons for their choice.

(7) Indicate whether all former vectors have been found and their levels of density (indoor resting, man-biting indoors and outdoors) as compared with those during the baseline observations and those obtained during the application of attack measures.

(8) Correlate the man-biting densities with the human activities outdoors at night and any increased exophilic tendency of vectors responsible for malaria transmission.

(9) Give results of dissection for infectivity and parity.

(10) Indicate whether there has been any change in the susceptibility levels of vectors to insecticides and consider the possible effect of any pesticides used in agriculture.

(11) Give conclusions as to the possible causes of resumption of transmission.

(12) Recommend action for immediate implementation of remedial control measures, taking into consideration the above-mentioned findings, and also any further protective measures required in the case of continued vulnerability or the presence of land development activities favouring a dramatic increase in vector output.

VIGILANCE

In areas under maintenance or where malaria eradication has been certified

Spot checks: reports should:

(1) Give the localities surveyed and the area they represent and the basis of their selection.

Annex 1

(2) Analyse the results following the outline given above for areas where house spraying has been withdrawn.

(3) Recommend localities where trend observations should be implemented on a seasonal basis.

Trend observations. Reports presented should:

(1) Describe the localities of trend observations, the areas they represent and their degree of vulnerability, indicating the localities for which data are available from the time of baseline observations until the discontinuation of attack measures.

(2) Indicate whether all vectors have been found during the period of observations, or, if one has disappeared, give the conclusive evidence as obtained by techniques and procedures given under section 3.5.2 of Chapter 3.

(3) Compare the density levels obtained in the present trend observations with those previously obtained from the same localities (or the nearest locality in the same area) and at the same season using the same techniques. Indicate whether vectors have regained their pre-spraying level of density and explain the prevailing environmental conditions that influenced differences in density levels.

(4) Give the density levels as obtained from localities other than those mentioned above in (3) and compare these with other localities observed in the present investigation and previously and explain the influence of the prevailing environmental conditions.

(5) Compare the human blood index of vectors as presently determined with those previously recorded and indicate whether changes in animal husbandry occurred that may have influenced changes in this index.

(6) Indicate whether there has been any change in the susceptibility level of vectors to insecticides and examine this in the light of any pesticides used in agriculture. Based on this, recommend the insecticide to be used for protective measures or in case of emergency.

(7) Give any additional observations that would support conclusions as to the potential vectorial efficiency of the predominant vectors, such as the parous rate or an alternative index of longevity.

(8) Derive from the above the overall potential efficiency of vectors for use in estimating the degree of receptivity of different areas represented by localities of the trend observations.

(9) Recommend protective measures and/or further studies to be undertaken on the basis of the above conclusions and in the light of:

(a) the degree of vulnerability and the determined level of receptivity,

(b) the influence of the present or planned land development and schemes for utilization or conservation of water.

BIBLIOGRAPHY

Alessandro, G. d' (1956) Aspects de l'anophélisme extra-domestique en rapport avec la lutte insecticide en Sicile, Rev. path. gén., 56, 186

Ariaratnam, V. & Brown, A. W. A. (1969) Exposure time versus concentration in the WHO standard test for mosquito resistance to chlorohydrocarbon insecticides, Bull. Wld Hlth Org., 40, 561-567

Beklemishev, W. N. (1957) Assessment of the density of vector populations in a malaria focus in relation to the evaluation of effects of imagicidal programmes. In: Beklemishev W. N. & Shipitsina N. V., eds, Seasonal phenomena in the life of malaria mosquitos in the USSR, Moscow, Medgiz., p. 462

Bertagna, P. (1959) Residual insecticides and the problem of sorption, Bull. Wld Hlth Org., 20, 861

Brown, A. W. A. & Pal, R. (1971) Insecticide resistance in arthropods, World Health Organization, Geneva

Chinaev, P. P. (1964) Exophilic mosquitos on virgin lands of the hunger steppe in central Fergana and in the delta of the Amu-Dariya river, Med. Parasitol, Moskva, 33, No. 5, 541-543

Cheng, F. Y. (1968) Responses of A. balabacensis to various patterns of DDT spraying of shelters in Sabah, East Malaysia, Bull. Wld Hlth Org., 38, 469-477

Cheong, W. H. et al. (1968) The incrimination of A. maculatus as the vector of human malaria at the Kelomitan-Thai border, Med. J. Malaya, 22, No. 3

Coz, J. et al. (1966) Etudes entomologiques sur la transmission du paludisme humain dans une zone de forêt humide dense, la région de Sassandra, République de Côte d'Ivoire, Cah. ORSTOM. sér. Ent. méd., IV, 7, 13-42

Cristesco, A. (1965) Contributions à l'étude de la composition par groupes d'âges des populations du groupe A. maculipennis par rapport à l'application des insecticides rémanents en Roumanie, Congrès nat. microbiol. méd., Bucarest, 208-209

Davidson, G. (1958) The practical implications of studies on insecticide resistance in anopheline mosquitos, Indian J. Malar, 12, 413

Davidson, G. & Jackson, C. E. (1961) DDT resistance in Anopheles stephensi, Bull. Wld Hlth Org., 25, 209

Detinova, T. S. (1962) Age-grouping methods in diptera of medical importance, Wld Hlth Org Monogr. Ser., 47

Detinova, T. S. & Gillies, M. T. (1964) Observations on the determination of the age composition and epidemiological importance of populations of A. gambiae Giles and A. funestus Giles in Tanganyika, Bull. Wld Hlth Org., 30, 23-28

Dow, R. P. et al. (1965) Dispersal of female Culex tarsalis into a larvicided area, Am. Journ. trop. Med. Hyg., 14, No.4., 656-670

Elliot, R. (1968) Studies of man-vector contact in some malarious areas in Colombia, Bull. Wld Hlth Org., 38, No.2, 239-253

Freyvogel, A. T. & Kihqule, P. M. (1968) Report on a limited anopheline survey at Hakara, South-Eastern Tanzania, Acta Tropica, 25, 17

Garrett-Jones, C. (1963) The human blood index of some anopheline mosquitos with reference to epidemiological assessment during malaria eradication WHO/MAL/371, (mimeographed unpublished document)

Gillies, M. T. & Wilkes, T. J. (1965) A study of the age-composition of populations of A. gambiae Giles and A. funestus Giles in North-Eastern Tanzania, Bull. Ent. Res., 56, 237

Gillies, M. T. & Wilkes, T. J. (1969) A comparison of the range of attraction of animal baits and of carbon dioxide for some West African mosquitos, Bulletin of Entomological Research, 59, Part 3, pp. 441-456

Gillies, M. T. & Wilkes, T. J. (1972) The range of attraction of animal baits and carbon dioxide for mosquitoes. Studies in a freshwater area of West Africa, Bulletin of Entomological Research, 61, Part 3, pp. 389-404

Gramiccia, G. (1956) Anopheles claviger in the Middle East, Bull. Wld Hlth Org., 15, 816

Gramiccia, G. Meillon, B. de, Petrides, J. & Ulbich, A. M. (1958) A. stephensi resistance to DDT in Southern Iraq, Bull. Wld Hlth Org., 19, 1102

Guy, Y. & Holstein, M. (1968) Données récentes sur les anophèles du Maghreb, Arch. Inst. Pasteur d'Algérie, 46, 142

Hamon, J. et al. (1965) Données récentes concernant la lutte contre les moustiques et les simulies, Médecine tropicale, 25, No.1, pp. 21-40

Keppler, W. J. et al. (1964) Reversion of Dieldrin resistance in A. albimanus Wiedmann, Mosquito News, 24, 1, 64-66

Khudukin, N. I. & Lisova, A. I. (1927) The possibility of winter infections with malaria, Med. Mysl Uzbek, 71, 6-7

Macdonald, G. (1973) Dynamics of tropical disease, Oxford

Macdonald, G & Göckel, C. W. (1964) Bull. Wld Hlth Org., 31, 365

Mattingly, P. F. (1962) Mosquito behaviour in relation to disease eradication programmes, Ann. Rev. Ent., 7, 419

Moshkovsky, S. D. (1946) The dependence upon temperature of the speed of development of malaria plasmodia in the mosquito, Med. Parazit. Mosk., 15, 19

Muirhead-Thomson, R. C. (1960) The significance of irritability, behaviourisitic avoidance and allied phenomena in malaria eradication, Bull. Wld Hlth Org., 22, 721

Pringle, G. (1966) A quantitative study of naturally acquired malaria infections in Anopheles gambiae and A. funestus in a highly malarious area of East Africa, Trans. roy. Soc. trop. Med. Hyg., 60, 626

Pringle, G. & Avery-Jones, S. (1966) An assessment of the sporozoite inoculation rate as a measure of malaria transmission in the Ubembe area of North-East Tanzania, J. Trop. Med. Hyg., 69, 132

Schlein, Y. & Gratz, N. G. (1974) Age determination of some anopheline mosquitos by daily growth layers of skeletal apodemes, Bull. Wld Hlth Org., 49, 371-375

Sloof, R. & Verdrager, J. (1972) Anopheles balabacensis, balabacensis Baisas 1936 and malaria transmission in South-Eastern areas of Asia, WHO/MAL/72.765 (mimeographed unpublished document)

Smith, A. (1962) Studies on domestic habits of A. gambiae that effect its vulnerability to insecticides, East Afr. Med. J., 39, 15

Smith, A. (1964) Studies on A. gambiae Giles and malaria transmission in the Umbugwe area of Tanganyika, Bull. Ent. Res., 55, 125

Smith, A., Obudho, W. O. & Esozed S. (1966) Resting patterns of A. gambiae in experimental huts treated with malathion, Trans. roy. Soc. trop. Med. Hyg., 60, 401

Spencer, M. (1974) A simple field method for assessing the proportion of older parous females in an anopheline sample (Anopheles farauti), Trans. roy. soc. trop. Med. Hyg., 68, 15

Thomas, V. (1965) Effects of certain extrinsic and intrinsic factors on the susceptibility of larvae of C. pipiens fatigans Wied to DDT, Mosquito News, 25, 38

WHO Expert Committee on Insecticides (1970) Seventeenth Report, Geneva, Wld Hlth Org. techn. Rep. Ser., No. 443

GENERAL BIBLIOGRAPHY

Boyd, M. F. (Ed.) (1949) Malariology. Philadelphia, Saunders, 2 vols.

Clark, L. R., Geier, P. W., Hughes, R. D. & Morris, R. F. (1967) The ecology of insect populations in theory and practice. London, Methuen

MacArthur, R. & Connell, J. (1966) The biology of populations. New York, Wiley

Macdonald, G. (1957) The epidemiology and control of malaria. London, Oxford University Press

Muirhead-Thomson, R. C. (1968) Ecology of insect vector populations. London, Academic Press

Pampana, E. (1962) A textbook of malaria eradication. 2nd Ed. London, Oxford University Press

Prothero, R. M. (1965) Migrants and malaria. London, Longmans

Russel, P. F., West, L. S., Manwell, R. D. & Macdonald, G. (1963) Practical malariology (2nd Ed.). London, Oxford University Press

Southwood, T. R. E. (1966) Ecological methods. London, Methuen

Whittaker, R. H. (1970) Communities and ecosystems. New York, Macmillan

WHO Expert Committee on Malaria, Eleventh Report, Geneva (Wld Hlth Org. techn. Rep. Ser., 1964, No. 291)